THE INTERPLANETARY CLOUD

✦

From AI Disruption and Planetary Limits to New Jobs, New Resources and an Abundant Future Beyond Earth

RISHABH KUMAR JAIN

COPYRIGHT

This book discusses artificial intelligence, data centres, energy, climate, employment, robotics, communications, space exploration and future infrastructure. It combines established science and engineering, emerging technologies, informed technological extrapolation, and the author's own speculative proposals.

Where existing technologies or scientific principles are discussed, they are presented according to current knowledge available at the time of writing. Future concepts—including large-scale extraterrestrial data centres, interplanetary backup systems, autonomous planetary industry, space-based communication and computing nodes, and other proposed infrastructure—should be understood as forward-looking concepts rather than claims that such systems currently exist or are certain to be built.

First Edition, 2026

ISBN: 978-1-7648938-2-4

Written by Rishabh Kumar Jain

ACKNOWLEDGEMENTS & DEDICATION

In loving memory of my parents, whose love, values and sacrifices laid the foundation for everything I have achieved; and to my siblings, nieces, nephew and in-laws, whose love and encouragement continue to support me. This book is dedicated to you all.

This book also comes from a period of extraordinary change in my own profession. Having spent years working across systems engineering, cloud computing, DevOps, Site Reliability Engineering, platform engineering and artificial intelligence, I have watched AI move rapidly from an experimental technology into a tool capable of performing work that technology professionals themselves once performed manually.

That experience made the subject of this book personal. Like millions of workers, I have had to ask what my profession may look like when machines can increasingly generate code, analyse failures, automate infrastructure and perform tasks that previously required significant human effort. My response has been to continue learning, to work with AI rather than simply compete against it, and to think beyond the immediate disruption toward the industries and opportunities this technology might eventually create.

I am grateful to the engineers, scientists, researchers, space agencies, technology companies and generations of thinkers whose work has made it possible even to ask the questions explored in this book. Most importantly, I thank the reader for being willing to look beyond both extremes of the AI debate—the belief that AI will automatically solve humanity's problems and the fear that it will inevitably destroy our future. There may be another path.

A NOTE ON EVIDENCE, FORECASTING AND SPECULATION

This book deliberately crosses several disciplines: artificial intelligence, software engineering, data-centre architecture, electrical power, climate and environmental constraints, telecommunications, orbital mechanics, robotics and planetary exploration. Because it moves from technologies that already exist to infrastructure that may not exist for decades or centuries, establishing a clear boundary between fact and extrapolation is essential. Throughout this book, three levels of discussion are used. What exists today includes technologies and scientific principles already demonstrated or operational: modern AI systems, terrestrial data centres, cloud computing, solar power, robotic planetary exploration, reusable rockets, the International Space Station, radio communication with spacecraft, laser-communication experiments and established orbital mechanics.

What is emerging includes technologies under active research or early development that could become building blocks for much larger systems: increasingly autonomous robots, AI agents, optical space communications, reusable launch systems, lunar resource utilisation, advanced solar power, robotic construction and more autonomous spacecraft.

What this book proposes goes considerably further. It asks what could happen if humanity combines these technologies over decades and centuries to construct data centres beyond Earth, establish intermediate computing and communication nodes through the Solar System, mine extraterrestrial resources robotically, maintain interplanetary copies of important information and eventually build an infrastructure in which humans, AI and robots expand together beyond Earth. These proposals are not predictions with fixed dates. Some may prove practical, others may require technologies not yet invented, and some may ultimately prove uneconomical or impossible. The purpose is to ask a larger engineering question:

If AI continues to grow while Earth's environmental, energy, land and resource constraints also increase, what infrastructure would humanity eventually need to allow both AI and civilisation to flourish without sacrificing the planet that made them possible?

READER'S NOTE

No background in artificial intelligence, computer science, data centres, networking, electrical engineering or space science is required to read this book. Technical terminology is retained where it helps explain an important idea, but every important technical term is explained in simple language when it first appears. For example, latency simply means the delay between sending information and receiving it. Replication means maintaining another copy of data somewhere else. Orbital velocity is the sideways speed an object needs to keep falling around a planet rather than falling directly onto it. The purpose is not to turn the reader into an AI engineer, network architect or rocket scientist. It is to provide enough understanding to follow the larger argument.

This book can broadly be read in three stages. The first examines the present: AI's effect on employment, software engineering, professional work, data centres, electricity consumption, heat, land, climate and public resistance. The second examines the transition: AI agents, robotics, autonomous systems, communications, renewable power and the technologies humanity is already developing that could form the foundations of something much larger. The third examines the future: the author's proposed progression toward orbital and planetary data centres, autonomous extraterrestrial industry, interplanetary communication and storage networks, robotic resource extraction, human-support stations and eventually a Solar System infrastructure capable of supporting both biological and artificial intelligence. The further the book travels into this third stage, the more speculative it becomes. That is intentional. The aim is not to tell readers that this future will happen. It is to investigate how it could happen.

Because interplanetary infrastructure sits at the intersection of network architecture, orbital physics, and physical logistics, certain foundational constraints—such as store-and-forward delay and propagation limits—are deliberately re-examined across Part VI through these distinct operational lenses.

PREFACE

Why I Am Writing This Book

Artificial intelligence presents humanity with a strange contradiction. We have created a technology capable of helping us solve increasingly difficult problems, yet many people now fear that the same technology may reduce the need for their own work. At the same time, the physical infrastructure supporting AI is expanding rapidly. Behind every apparently weightless digital interaction are processors, data centres, electricity, cooling systems, networks, land, water and materials.

For me, the first part of this contradiction is personal.

My professional background is in IT infrastructure, cloud computing, DevOps, Site Reliability Engineering and platform engineering. I have spent years learning how to build, automate and maintain the digital systems on which modern organisations depend. Now I am watching artificial intelligence learn to assist with—and increasingly perform—some of the same work.

That experience initially raises an uncomfortable question: **If AI can eventually perform more of what people do, what happens to the people?**

But while thinking about that question, I encountered another.

If AI continues expanding across science, medicine, engineering, transport, manufacturing, robotics, communications and almost every other major industry, **where will all of the physical infrastructure supporting that intelligence go?**

AI may appear to live inside our computers and phones, but intelligence at this scale has a physical body. Data centres require enormous amounts of electricity. Their processors generate heat that must be removed. Facilities require land, cooling infrastructure, networks, backup systems and increasingly sophisticated power systems. Their construction also depends on metals, minerals and global supply chains.

Earth must provide all of this while continuing to support billions of people, cities, agriculture, industry, ecosystems and the natural environment.

These two problems—**the future of human work and the physical limits of computing on Earth**—may appear unrelated. This book asks whether they could eventually become parts of the same solution.

Much of the discussion around AI employment focuses on how society might adapt if machines allow organisations to operate with fewer people. That response is necessary, but it may be incomplete. What if the longer-term objective should not simply be to divide a shrinking amount of human work among more people? What if AI and automation allow civilisation to attempt things that were previously too expensive, dangerous or complicated to undertake at scale?

That possibility led me to what I call the **Employment Bridge**.

The Employment Bridge does not assume that every existing job can or should be preserved. Technologies have always changed occupations. Instead, it asks how we move from a period in which AI automates existing tasks toward one in which the capabilities of AI help create entirely new industries, infrastructure and forms of human work.

At the same time, I propose another principle: **Responsible Acceleration**.

Technological progress should not require us to sacrifice the planet or destabilise societies in order to reach some imagined future more quickly. Earth comes first. Computing should become more efficient. Energy systems should become cleaner and more resilient. Water, land and other resources should be used responsibly. Workers need opportunities to adapt, learn and participate in the industries being created around them.

But protecting Earth does not necessarily mean that civilisation must permanently limit its ambitions to the physical resources available on Earth's surface.

Humanity has another frontier.

Beyond our atmosphere lies an enormous Solar System containing energy, materials and environments in which machines can potentially operate without competing directly with cities, farmland and terrestrial ecosystems. We already operate spacecraft autonomously, maintain infrastructure in orbit, communicate across planetary distances, land robots on other worlds and return increasingly sophisticated machines to space.

These capabilities are primitive compared with what would be required for an interplanetary industrial civilisation. But they raise an important possibility.

What if, over decades and centuries, humanity gradually extended some of its digital and industrial infrastructure beyond Earth?

Data centres could begin with specialised orbital systems. The Moon could eventually support computing, communications, storage, science and industry. Mars and more distant locations could support workloads that do not require immediate communication with Earth. Autonomous machines could construct infrastructure, extract local resources and perform work in environments dangerous or impractical for humans. Humans would continue to provide judgement, objectives, governance, engineering and, where appropriate, physical presence.

As that infrastructure expanded, something even more important could happen.

New infrastructure creates new industries. New industries create new occupations. Space logistics, robotics, energy systems, communications, autonomous mining, construction, maintenance, planetary networking, AI safety, scientific research and countless activities we cannot yet name could create forms of meaningful human work that do not exist today.

And the resources available to civilisation would no longer be defined entirely by what can be extracted from one inhabited planet.

This is where the argument becomes larger than data centres.

Many of humanity's persistent problems are intensified by scarcity. Nations compete for energy, resources, industries and economic advantage. People compete for limited employment opportunities. Economic shocks can spread through deeply interconnected systems. Poverty persists despite enormous technological progress, while access to resources and opportunity remains profoundly unequal.

Expanding civilisation beyond one planet would not automatically eliminate poverty, unemployment, inequality or financial crises. These are also problems of institutions, politics, distribution and human behaviour. But a civilisation with access to vastly greater resources, energy, productive capacity and technological capability would operate under very different constraints from one confined permanently to a single planetary economy.

AI could therefore represent something much larger than the automation of today's jobs.

It could help us **create more things worth doing**.

Instead of asking only how many existing jobs AI might remove, we can also ask how many new industries greater intelligence could make possible. Instead of asking only how many more data centres Earth can accommodate, we can ask whether computing infrastructure must always remain confined to Earth. Instead of assuming that humanity must forever compete over the resources accessible on one planet, we can investigate how responsibly expanding our technological reach might enlarge the resource base available to future generations.

This is the central journey of *The Interplanetary Cloud*.

It begins with the very immediate fear of AI disruption. It then confronts the environmental and physical consequences of expanding digital intelligence. From there, it explores how computing, robotics, energy, communications and industry might gradually extend into orbit, onto the Moon, to Mars and eventually farther across the Solar System.

The destination is not a future in which machines replace humanity.

It is a future in which **humanity and machines expand what civilisation is capable of doing together**.

The Interplanetary Cloud is therefore not simply a proposal for putting data centres in space. It is a thought experiment about changing the scale of the problem. If AI gives civilisation unprecedented intelligence and automation, perhaps our response should not be to accept a future of fewer opportunities within the same physical boundaries. Perhaps we should use those capabilities to enlarge the boundaries themselves.

That transition would take generations. Some ideas explored in this book may prove practical; others may evolve into forms we cannot currently anticipate. The scientific and engineering constraints discussed throughout the book remain real. Nothing about expanding beyond Earth removes the need to protect the planet on which civilisation began.

Quite the opposite.

Earth is the foundation from which everything else must be built.

The immediate task is therefore to protect the present while preparing for a larger future: use AI responsibly, reduce its environmental burden, help people adapt, develop new industries, improve robotics and energy systems, and gradually build the infrastructure that could allow civilisation to operate beyond one world.

If we succeed, the story of artificial intelligence may ultimately be very different from the one many people fear today.

It may not be the story of humanity creating machines that leave people with less to do and fewer resources to share.

It may become the story of humanity creating intelligence powerful enough to help us **build more, explore farther, create new work, access new resources, restore our planet and expand the possibilities available to future generations.**

That is why I am writing this book.

Contents

PART I
THE DISRUPTION

When automation reaches the people who built automation

What PART I Covers

This Part begins with the disruption already happening around us. Artificial intelligence is moving from assistant to agent: from generating answers to performing tasks, coordinating with other systems and changing the nature of professional work. The chapters examine what this means for engineers and other knowledge workers, why human judgement and oversight remain important, and how automation may change rather than simply eliminate occupations. But AI is not only a labour story. Every model ultimately depends on processors, electricity, cooling, networks and data centres. The Part therefore ends by moving from the visible disruption of work to the less visible physical infrastructure required to support increasingly capable artificial intelligence.

Chapter 1

AI Came for the People Who Built IT

For decades, one of the safest pieces of career advice in technology was simple: **keep learning**. Technology would change, old tools would disappear and new ones would replace them, but people who continuously updated their skills could move with the industry. System administrators learned virtualisation. Network engineers learned automation. Infrastructure teams moved into cloud computing. Software developers learned new programming languages and frameworks. Operations teams adopted **DevOps**, a way of bringing software development and IT operations closer together so applications can be released and maintained faster. Then came containers, Kubernetes, Infrastructure as Code and increasingly automated deployment pipelines. Containers package an application together with what it needs to run so that it behaves consistently across different computers. Kubernetes is a platform that automatically deploys and manages large numbers of those containers. Infrastructure as Code (IaC) means defining servers, networks and cloud resources in configuration files rather than building them manually. CI/CD, short for Continuous Integration and Continuous Delivery, automates the testing and release of software changes.

Each generation of automation removed some manual work, but it also created another layer for technology professionals to learn. The expectation was that automation would remove repetitive tasks while humans moved upward into architecture, engineering, reliability, security and increasingly complex problem-solving. Artificial intelligence is beginning to challenge that comfortable assumption because, for the first time, the automation is not limited to one clearly defined technical process. Generative AI can write code, generate configuration, interpret an error message, summarise an incident, explain unfamiliar software, draft documentation, analyse logs and propose an architecture. Increasingly, AI agents can also interact with tools and perform sequences of tasks rather than simply answering questions.

The International Labour Organization's research suggests that this change should not yet be described as the wholesale replacement of human workers. Its 2025 global analysis found that about one in four jobs had some exposure to generative AI but concluded that **transformation of jobs was more likely than complete replacement**. Its 2026 review of empirical evidence similarly found that large-scale employment displacement remained limited so far, while warning about pressure on younger workers and changes to how work is organised. (International Labour Organization) That distinction matters. AI does not have to eliminate an occupation to disrupt the person doing it.

I Am Experiencing This Transition Too

I am not observing this change only as an author writing about the future. I am experiencing it from inside the technology industry. My career has included systems engineering, cloud infrastructure, DevOps, Site Reliability Engineering and platform engineering. Site Reliability Engineering (SRE) applies software-engineering methods to keeping large computer systems reliable and available. Platform engineering involves building shared infrastructure and tools that make it easier for other engineering teams to develop and operate applications. Like many engineers, I adapted repeatedly as the industry changed. I learned cloud platforms, automation, Infrastructure as Code, containers, Kubernetes, security and other technologies because staying current was part of remaining employable. The arrival of generative AI changed the nature of that adaptation.

Tasks that previously required an engineer to search through documentation, compare examples, experiment with syntax and spend considerable time troubleshooting can increasingly begin with a conversation with an AI system. An engineer can describe the infrastructure required and ask AI to draft a Terraform configuration. **Terraform** is a tool that allows engineers to describe cloud infrastructure in configuration files and automatically create it. AI can analyse a Kubernetes error, propose a deployment configuration, explain a networking problem or generate the first version of an automation script. That creates an uncomfortable question for someone who has spent years accumulating technical knowledge: If AI can produce in seconds something that took me years to learn how to produce, what exactly is the value of my experience? I do not think the answer is that my experience has become worthless. In many ways, the opposite has happened. AI makes technical knowledge more accessible, but accessibility is not the same as judgement.

Suppose I ask an AI system to create infrastructure for a production application. It may produce perfectly valid configuration. But is the architecture secure? Is it unnecessarily expensive? What happens when one region fails? Where are the secrets stored? Can the application scale? Are the network permissions too broad? What happens during an upgrade? Can the change be rolled back safely? AI can generate an answer without carrying responsibility for the consequences. The experienced engineer increasingly provides that missing context. Yet there is another side to this. If an experienced engineer using AI can accomplish significantly more work, a company may decide that it needs fewer engineers to produce the same output. The disruption can therefore happen even if AI never becomes capable of independently replacing an engineer. The equation may change from:

10 engineers × traditional tools

to something closer to:

6 engineers × AI-assisted tools = similar or greater output.

The exact numbers will vary enormously between organisations, and this is not a prediction that four out of every ten engineers will lose their jobs. It simply demonstrates the economic mechanism. Productivity gains can create new work, lower costs and increase output, but they can also allow organisations to operate with smaller teams. Research into real AI coding use already shows why software professionals are watching this carefully. Anthropic's analysis of 500,000 coding interactions found that use of its coding agent leaned strongly toward automation rather than simple assistance, while more recent research shows coding agents taking on increasingly complex tasks. (Anthropic)

How I Am Trying to Adapt

My response has not been to compete with AI at the task it performs best. Trying to prove that I can manually write a configuration file faster than an AI system would make little sense. Instead, I have been trying to understand how to work with AI while strengthening the skills required to judge, integrate and govern what it produces. That means using AI to accelerate research, troubleshooting, drafting and technical exploration while continuing to understand the underlying systems. It also means learning more about AI itself rather than treating it as somebody else's technology. This is an important distinction. An engineer who simply copies an AI-generated Kubernetes configuration may become more dependent on AI. An engineer who understands Kubernetes and uses AI to generate the first draft can become more productive. The same applies to software development. AI can write a function, but the developer still needs to understand what the application is supposed to do, how different components interact, how data should be protected and what happens when the generated code encounters conditions that were never included in the prompt.

My own strategy has therefore become less about collecting individual commands and more about strengthening **systems thinking**—understanding how networking, cloud infrastructure, security, applications, automation, data and increasingly AI fit together. That is also one reason this book exists. Rather than seeing AI only as something that threatens today's technology jobs, I want to ask what new infrastructure, industries and professions increasingly capable AI could eventually help humanity create.

Software Engineers Are at the Front of the Change

Software engineering is one of the clearest laboratories for understanding what may happen elsewhere. A traditional developer receives requirements, designs a solution, writes code, tests it, reviews changes and fixes defects. AI can now participate in almost every one of those stages. A developer might describe a feature in ordinary English and receive a first version of the code. AI can write unit tests, which are small automated checks that verify whether individual pieces of software behave correctly. It can review code for possible errors, explain an unfamiliar codebase, suggest database queries and generate documentation. This does not necessarily mean software developers disappear. What may change is that developer increasingly becomes responsible for describing problems clearly, designing architecture, integrating systems, evaluating AI-generated code and making decisions that require knowledge of the wider business.

In that sense, software engineering may provide an early preview of AI's effect on many professional occupations: automation moves upward from individual repetitive tasks toward portions of knowledge work.

The Problem for Junior Workers

There is, however, a less obvious risk. Senior software engineers became senior engineers by first being junior engineers. They wrote small functions. They fixed relatively simple bugs. They read documentation. They made mistakes. They reviewed other people's code. They gradually encountered larger systems and more serious failures. AI is particularly good at many of those smaller tasks. That creates what we might call the apprenticeship problem. If organisations decide that a senior developer with AI can perform the work previously assigned to several junior developers, where will the next generation of senior developers come from? The same issue exists in infrastructure engineering. Engineers learn troubleshooting partly by troubleshooting. If an AI system immediately diagnoses every beginner-level problem, the junior engineer may solve incidents faster while gaining less understanding of why the system failed. The answer cannot simply be to ban AI from junior workers. That would prepare them for a workplace that no longer exists. Instead, training may have to change so that workers are deliberately taught to challenge AI answers, investigate underlying mechanisms and solve selected problems without automation.

Learning how to use AI and learning how systems work cannot become opposing choices. We need both.

The Disruption Is Not Limited to Technology

It would also be a mistake to write this chapter as though DevOps engineers and software developers are the main people facing this transition. The effects extend much further. An administrative worker can use AI to draft correspondence, summarise meetings, prepare reports and organise information. A marketing professional can generate campaign ideas, social-media content and first drafts of advertising copy. An accountant can use AI-assisted systems to classify information, investigate transactions and prepare explanations. Customer-service staff increasingly work alongside AI systems that retrieve information and draft responses. Human-resources teams can use AI to prepare job descriptions, summarise policies and analyse employee information. Lawyers can use AI to search and summarise large collections of documents, although professional judgement and verification remain essential. Designers can generate visual concepts rapidly. Translators can begin with machine-generated translations. Researchers can use AI to summarise literature and explore unfamiliar subjects. Clerical occupations are currently among those with the highest measured exposure to generative AI, while exposure is also increasing in professional and technical occupations such as financial analysis, web development and application programming.

This means the AI employment question is not simply: Will programmers lose their jobs? It is: What happens when a large part of the knowledge economy receives a tool capable of performing portions of knowledge work?

A Job Is Really a Collection of Tasks

This is where the discussion becomes more useful. We normally think in terms of occupations: accountant, engineer, lawyer, teacher, administrator, designer. AI does not necessarily automate an occupation. It automates tasks inside the occupation. Consider a hypothetical job containing ten regular activities. AI may perform three almost completely, assist with four, and have little ability to perform the remaining three. The job still exists. But it is no longer the same job. Over time, the organisation may redesign it around the remaining tasks. Those tasks are likely to involve judgement, accountability, interpersonal communication, physical interaction, unusual situations or deep domain knowledge. This task-level view is also why predictions that millions of jobs will either simply "vanish" or remain completely unaffected should be treated carefully. The ILO's work specifically analyses occupations through their component tasks because exposure varies significantly even within the same profession. (International Labour Organization)

AI Can Also Expand What One Person Can Do

There is another side that deserves equal attention. AI can allow people to perform work that previously sat outside their expertise. A systems engineer can get help understanding application code. A software developer can explore infrastructure. A small-business owner can produce marketing material without immediately hiring an agency. Someone with an idea for a simple application may be able to prototype it without first assembling an entire development team. If AI only allows companies to produce the same things with fewer workers, the employment story becomes primarily one of displacement. If AI also allows individuals and organisations to attempt things that previously cost too much, new demand can emerge. And that is where this book begins to look beyond the immediate employment debate.

The Number of Possible Jobs Is Not Fixed

Imagine explaining the profession of cloud engineer to someone in 1980. The person might understand computers and telecommunications but struggle with the idea that companies would rent enormous pools of computing infrastructure through the Internet and employ specialists to automate those resources. Likewise, today's economy contains cybersecurity analysts, mobile-application developers, cloud architects, social-media managers and many other professions that either did not exist or existed only in small, specialised forms several decades ago. None of this guarantees that future job creation will occur quickly enough or fairly enough to compensate everyone affected by automation. That transition problem is real, and we will return to it later. But it suggests that asking only, "Which existing jobs will AI remove?" gives us only half of the picture. We should also ask: What will humanity attempt because AI makes previously impossible projects possible? That question eventually takes this book far beyond the office.

From Automation to Expansion

AI could help design better materials, manage electricity grids, accelerate scientific research, coordinate robots, optimise spacecraft, operate remote infrastructure and control machines in environments where humans cannot safely live. If those capabilities continue improving, AI may eventually help humanity expand its industrial activity beyond Earth. That changes the employment question. A Platform Engineer worried today about AI-generated infrastructure code may one day work in an industry managing autonomous infrastructure distributed between Earth and orbit. A network engineer's knowledge may evolve toward communication systems spanning planetary distances. A reliability engineer may supervise machines that must operate for months without direct human maintenance. These particular job titles are speculative. The larger principle is not. Technology can eliminate tasks while simultaneously expanding the boundary of what civilisation is capable of doing. The challenge is surviving the transition between those two realities.

For someone experiencing AI disruption today, being told that entirely new industries may exist decades from now is not enough. People need income, dignity, education and opportunity now. Companies need to think beyond short-term head-count reduction. Governments and education systems need to reconsider how people enter professions when the traditional beginner tasks are increasingly automated. At the same time, we should resist assuming that today's economy represents the final catalogue of useful human work. AI is forcing us to reconsider what a worker does. Eventually, it may force us to reconsider where civilisation itself does its work. Before we can reach that future, however, we need to understand something frequently hidden by the word AI. Artificial intelligence is not merely software. Every AI model ultimately runs on physical machines. Those machines require electricity. They generate heat. They require land, networks, materials and increasingly enormous data centres. And that brings us to the second disruption explored in this book: AI may be changing human work, but it is also beginning to change the physical infrastructure of the planet.

Chapter 2

From Copilot to Coworker

The first wave of generative artificial intelligence entered most workplaces as an assistant. A person asked a question and AI answered it. A developer requested code, an administrator asked for an email, or an employee requested an explanation. The human remained clearly in control because the AI generally waited for the next instruction.

That model is evolving. The next stage is **agentic AI**, in which an AI system can pursue a goal, use authorised tools, examine the result of an action and decide what to do next. Instead of merely telling a worker how to perform a task, AI increasingly participates in performing it. This shift from **assistance to delegated action** could become one of the most important changes in workplace automation.

From Answering a Question to Performing a Task

Suppose an engineer tells an AI assistant: "Our application is returning an error when users try to log in. What should I investigate?" A conventional assistant might recommend checking application logs, network connectivity, authentication settings and recent deployments. The engineer then performs those checks. An authorised AI agent could go further. It might retrieve the logs itself, inspect recent deployments, examine monitoring information, compare the failure with earlier incidents and report the likely cause. It might then prepare a correction for human approval. The difference is simple but important:

AI assistant → recommends what to do

AI agent → performs authorised steps toward the goal

To do this, the agent needs **tools**. A tool is an authorised function that allows AI to interact with another system—for example, reading monitoring data, querying a database, creating a support ticket or calling a cloud-management API. An **API (Application Programming Interface)** is a controlled way for software systems to communicate with one another. The basic architecture therefore becomes:

Human Goal → AI Agent → Authorised Tools → Real Systems

The human supplies the objective while the agent determines some of the intermediate steps.

Automation Is Not New—But Agents Are Different

Companies have automated repetitive processes for decades. Scheduled backups copy files automatically. Monitoring systems generate alerts when servers fail. Payroll systems calculate salaries according to predefined rules. Traditional automation commonly follows something resembling:

IF this happens → THEN perform this action.

An AI agent can operate more flexibly. It can receive a broader objective, examine information, select among available tools, evaluate what happened and decide what step should follow. This does not make the agent unrestricted or infallible. In fact, the greater its ability to act, the more important its boundaries become.

An AI Should Not Have Every Key

Modern computer systems use **Role-Based Access Control (RBAC)** to limit what users and software are permitted to do. AI agents need similar restrictions. A diagnostic agent might be allowed to read logs but not delete them. Another agent might restart an application but require human approval before changing important infrastructure. This follows the security principle of least privilege: give a user or system only the access required to perform its role.

The distinction matters because AI can be wrong. If an agent incorrectly concludes that a database is corrupted but has only read access, it produces a bad recommendation. Give the same agent unrestricted administrative access and the mistake could become a major production incident. The intelligence of an AI cannot therefore be separated from the authority we give it.

The Human Role Starts Moving Upward

Agentic AI also changes the employment question. A DevOps engineer might traditionally receive an alert, open monitoring systems, search logs, inspect

Kubernetes, check a recent deployment, communicate with developers and prepare a fix. If an AI agent performs much of the initial investigation automatically, the engineer's role begins moving from gathering information toward **validating conclusions, approving important actions and handling unusual situations**. The same principle can apply across many professions. AI may perform more routine sequences while humans concentrate on judgement, exceptions, accountability and interaction with other people. The progression can be summarised as:

Human performs every step → AI assists individual steps → AI performs routine sequences → Human supervises outcomes and exceptions.

This does not guarantee that every existing job will survive unchanged. Some tasks and positions may disappear while others evolve. But it suggests that one important human role will increasingly be supervising systems that perform work previously carried out manually.

But Humans Still Need to Learn

There is also a less obvious risk. Imagine an engineer who once investigated twenty incidents each month. Those incidents gradually taught the engineer how complex systems behave. If AI investigates eighteen automatically and gives the engineer only the two most difficult cases, productivity may improve—but where does the engineer gain the experience required to solve those difficult cases?

Organisations may therefore need to preserve **human learning loops** deliberately. Instead of an agent simply reporting *Problem resolved*, it could show the evidence it examined, the possibilities it rejected, the action it selected and why. Junior engineers could review automated investigations rather than being completely removed from them. AI should not only perform work. Where practical, it should help humans understand the work it performs.

When Agents Begin Talking to Agents

The next stage goes beyond one human working with one AI agent. Specialised agents can exchange information, use different tools and coordinate parts of a larger process. Machines may eventually perform some of the communication that currently passes repeatedly between departments, while humans supervise consequential decisions and unusual situations.

This becomes particularly important when AI interacts with physical infrastructure. Agents could monitor computing systems, coordinate robots, respond to equipment failures or help manage electrical networks. Eventually, similar architectures could operate where humans cannot immediately intervene—remote data centres, the Moon, Mars or infrastructure travelling through space.

But machine-to-machine coordination creates new problems. **What happens when one agent is wrong and other agents trust it? How much authority should they receive? When must humans intervene? And how can agents belonging to different organisations trust one another?** The next chapter explores these questions through a railway disruption in which one event must be understood and coordinated across an entire network of specialised AI agents.

Chapter 3

When Machines Start Talking to Machines

AI becomes much more consequential when it stops merely helping one person and starts coordinating with other AI systems. Large organisations already operate separate systems for monitoring, maintenance, scheduling, security and customer communication. When something unusual happens, humans often become the connection between them—someone sees an alarm, opens a ticket, contacts another team and explains what happened. Multi-agent AI could automate parts of this coordination. A multi-agent system consists of specialised AI agents that divide a larger problem into different responsibilities and exchange information to solve it together.

A Signal Fails During the Morning Rush

Imagine a busy railway during the morning rush. A signalling component develops a fault, and a monitoring system detects the problem. The railway control centre receives an alert, but responding to the incident requires much more than detecting it. Operations need to understand which trains are affected. Maintenance needs to know what equipment may have failed and which technicians are available. Timetable planners need to adjust services. Passenger-information systems need accurate updates. If the disruption becomes serious, replacement transport may also be required.

The technical fault may be detected within seconds, while the organisational response takes much longer because information must move between different systems, people and departments. A future multi-agent system could change that. An Operations Agent receives the alarm and identifies the affected location and trains. A Diagnostic Agent examines equipment telemetry and maintenance history. A Maintenance Agent checks technicians and replacement equipment. A Scheduling Agent examines alternative train movements, while a Passenger Information Agent prepares updates using confirmed information. The operational loop becomes:

Sensor → Operations Agent → Diagnostic Agent → Maintenance/Scheduling Agents → Passenger Agent → Human supervision

These agents do not necessarily replace railway controllers, engineers or technicians. Their immediate value is removing some of the repeated information gathering and communication surrounding those people.

Automation Can Multiply an Error

Speed also creates risk. Suppose the Operations Agent identifies the wrong section of track. Other agents trust that information. Maintenance personnel could be sent to the wrong location, trains could be diverted unnecessarily, and passengers could receive incorrect information. Humans can also spread incorrect information, but machines can propagate a mistake across multiple systems almost instantly. The same capability that makes multi-agent coordination powerful can therefore make errors spread faster. Critical systems need safeguards. Important observations may require confirmation from another sensor or source. Agents should distinguish between a measured fact and an AI interpretation. For example:

Sensor observation: communication with signalling controller lost.

AI interpretation: probable hardware failure. Those statements are not equivalent. Before a high-impact action is taken, the interpretation may need independent confirmation or human approval.

Humans Supervise Consequential Decisions

This changes the role of the human operator. Instead of spending much of an incident collecting information from different teams, a controller could receive a continuously updated operational picture assembled by the agents. Routine, low-risk actions might happen automatically. Consequential actions should receive stronger oversight. An agent restarting a passenger-information display is very different from an agent proposing to close an entire railway corridor. This creates an important principle: The greater the consequence of an automated decision, the stronger the human oversight should be. For some decisions, a human may supervise and intervene when necessary. For more serious decisions, explicit human approval may be required before the system proceeds.

Cybersecurity Becomes More Important, Not Less

Connecting agents to real infrastructure also increases cybersecurity risk. An AI responsible for passenger announcements should not have permission to alter railway signalling simply because both systems belong to the same organisation. The principle of least privilege introduced in the previous chapter therefore becomes even more important. Each agent should receive only the permissions necessary for its role. A Communications Agent communicates. A Diagnostic Agent investigates. A Scheduling Agent schedules. Safety-critical systems retain their own protected controls. This separation limits the damage that can result from either an AI mistake or a compromised component. Multi-agent architecture should not mean creating one enormously powerful AI with access to everything.

AI Will Not Always Know the Right Answer

Real incidents can involve conflicting priorities. An operational decision that improves the timetable might increase passenger overcrowding. A faster recovery action might introduce additional safety risk. Several agents may therefore recommend different actions. AI can analyse information and calculate possible outcomes, but some decisions involve judgement about safety, fairness, risk and consequences that cannot be reduced to a single optimisation target. This is where human responsibility remains particularly important. The basic multi-agent cycle becomes:

Observe → Interpret → Communicate → Coordinate → Act → Verify.

From Railways to Another Planet

There is a deeper reason for discussing railway automation in a book about the future of data centres and space. On Earth, when automation cannot resolve a problem, a human can usually reach the equipment eventually. An engineer can enter the data centre. A technician can inspect a railway signal. A mechanic can repair a machine. That assumption changes when infrastructure moves beyond Earth.

If equipment fails on the Moon, sending a technician is difficult. On Mars, real-time human control from Earth is impossible because communication itself takes minutes. Future extraterrestrial infrastructure will therefore need local systems capable of detecting problems, communicating with one another, coordinating robots and taking authorised actions without waiting for instructions from Earth.

The architecture is fundamentally similar to the railway example:

Something happens → specialised systems investigate → information is shared → resources are coordinated → authorised action is taken → humans supervise at the level that distance allows.

A future Martian facility might therefore use AI to coordinate mining equipment, energy systems, robots, communications and maintenance. The details are speculative, but the underlying direction is important: technology that appears today primarily as workplace automation may eventually become essential for operating infrastructure in environments where constant human intervention is impossible. And this creates the bridge to the next problem. Every AI agent requires computation. Computation requires physical machines. Those machines require electricity and produce heat.

Chapter 4

The Physical Cost of Intelligence: Data Centres, Energy, Heat and the Planet

Artificial intelligence often feels almost weightless. We type a question into a phone or computer and receive an answer seconds later. There is no visible factory, engine or power station. But AI does not exist in an abstract digital world. Behind every response are processors performing calculations, memory holding information, storage systems preserving data, networks moving information and cooling systems removing heat. These machines operate inside data centres connected to electricity grids, water systems and communication networks. **To the user, AI is software. To the planet, AI is infrastructure.**

AI Did Not Invent the Data Centre

Data centres existed long before generative AI. Banks, governments, telecommunications companies, universities, websites and cloud services have depended on them for decades. AI has not created this infrastructure model, but it is dramatically increasing its importance and computational demand. Two concepts explain much of that demand. Training is the computational process through which an AI model learns patterns from large quantities of data. Inference occurs when the trained model is used—for example, to answer a question, generate code, analyse an image or control another system. Training can require enormous bursts of computing power, while inference becomes significant because millions or billions of requests can accumulate across users and applications. As AI spreads into software development, offices, science, healthcare, robotics and industrial systems, the infrastructure supporting it must expand as well.

More Computation Means More Energy and Heat

Modern AI relies heavily on specialised processors such as **GPUs (Graphics Processing Units)** because they can perform many mathematical operations in parallel. Thousands of powerful processors operating together can perform extraordinary amounts of computation, but they also consume substantial electricity.

And electricity creates another problem: **heat**. A laptop becomes warm during demanding work. An AI data centre experiences the same physical effect on an enormous scale. Processors, memory, storage, networking equipment and power systems consume electricity, and much of that energy ultimately becomes heat. If this heat is not removed, equipment becomes less reliable and may eventually shut down. Cooling therefore becomes part of the computational infrastructure itself. Data centres can use air cooling, increasingly powerful liquid-cooling systems and other specialised techniques, but the basic physical relationship remains:

Electricity enters → computation occurs → heat is produced → heat must go somewhere.

Cooling does not destroy heat; it moves it somewhere else. This principle will become particularly important later when we consider data centres in space, where there is almost no surrounding air and unwanted heat must ultimately be released primarily through radiation.

The Electricity Problem

The scale is already significant. The International Energy Agency estimated that data centres consumed around **415 terawatt-hours of electricity globally in 2024**, about 1.5% of global electricity consumption, and projected that consumption could more than double to around 945 terawatt-hours by 2030 in its base case, with AI being the largest driver of the increase.

The global percentage tells only part of the story. Data centres cluster around locations with suitable land, electricity, network connectivity and customers. A country may have enough electricity overall while a particular city or electricity network struggles to connect several enormous new facilities.

A hyperscale data centre cannot simply be plugged into an ordinary electricity socket. Large facilities can require new substations, transmission infrastructure, transformers and additional generating capacity. The question therefore becomes not only: How much electricity does AI require? but: Where will that electricity come from, and can the grid deliver it reliably without creating unacceptable costs elsewhere? Renewable energy will be an important part of the answer. Solar, wind, storage, nuclear power and other technologies can contribute to different locations. But electricity supply must also match demand over time. The Sun sets, wind changes and batteries have finite capacity. AI infrastructure therefore becomes inseparable from energy infrastructure.

Water, Land and Communities

Electricity is not the only constraint. Some cooling systems can use significant amounts of water directly or indirectly, which becomes particularly sensitive in regions already experiencing water stress. Large data-centre campuses also require land, electrical equipment, cooling infrastructure, network connections, security areas and access roads.

This creates competition for resources. Land near growing cities may also be required for housing, industry, transport, agriculture or public infrastructure. Communities may question water consumption, new transmission infrastructure, noise, construction or possible pressure on electricity systems. Such concerns should not automatically be dismissed as opposition to technology. A data centre may create investment, employment, tax revenue and valuable digital infrastructure while still imposing costs on its surrounding community. Governments and technology companies therefore face a difficult balancing problem:

How much physical infrastructure should society dedicate to an increasingly computational economy, and who carries its environmental and economic costs? There is no environmentally neutral data centre. Every design involves trade-offs among energy, water, land, materials, cooling, reliability and cost.

Efficiency Helps, but Demand Can Grow Faster

AI processors will become more efficient. Cooling will improve. Software will use hardware more effectively, and smaller models may accomplish tasks that once required much larger systems. These improvements are essential. But efficiency does not automatically mean lower total consumption. If a future processor uses one-tenth as much electricity for the same AI task but society performs one hundred times as many AI tasks, total electricity demand can still increase. This is related to the Jevons paradox: making a resource more efficient to use can sometimes encourage so much additional use that overall consumption continues growing. Efficiency should therefore be our first line of defence, but it may not be the final answer.

Protect Earth First

This book will eventually explore a much more ambitious possibility: moving portions of civilisation's computing infrastructure beyond Earth. But space must not become an excuse for ignoring problems here. For the foreseeable future, Earth will remain the centre of human computing. We should use cleaner energy, improve processors and cooling, recycle water where appropriate, reuse waste heat where practical, extend hardware life, recycle valuable materials, locate facilities intelligently and avoid unnecessary computation. Not every problem requires the largest possible AI model.

The argument of this book is therefore not: Earth is becoming inconvenient, so move AI into space. It is: Protect Earth first, while developing technologies that may eventually prevent one planet from carrying civilisation's entire computational burden. That distinction is central to everything that follows. AI is still at an early stage. If its capabilities and use continue expanding over decades and centuries, humanity may eventually need far more computing capacity than today's terrestrial infrastructure can sustainably provide. At that point, the vast energy, resources and physical space beyond Earth become relevant—not as an escape from environmental responsibility, but as a possible extension of it.

Before we can seriously consider that future, however, we first need to understand the environment in which extraterrestrial infrastructure would operate. Space is not an empty warehouse where we can simply place servers. Everything is moving, energy and heat still obey physical laws, and every data centre, communication node and spacecraft would have to operate within the mechanics of the Solar System.

PART II
AI HAS A PHYSICAL BODY

The cloud is software to the user, but industry to the planet

Closing Part I

Part I began with people and ended with infrastructure. AI can automate tasks, coordinate systems and increase the productivity of individual workers, but none of those capabilities exist without physical machines consuming electricity and producing heat. The employment question and the infrastructure question are therefore connected. The next Part examines that physical foundation more closely: the resources, energy, cooling, land and industrial systems required to support an increasingly computational civilisation.

What Part II Covers

Part I established that AI is not merely software. This Part examines the physical consequences of that fact. Growing computation requires electricity, cooling, materials, land, networks and increasingly large data-centre infrastructure. Reliability adds further requirements because critical computing cannot simply stop when energy becomes scarce or equipment fails. The chapters examine resource competition, the unavoidable problem of heat, the collision between data-centre growth and surrounding infrastructure, and the wider question of how much of Earth's physical capacity should be dedicated to computation. The central argument is not that AI growth must stop, but that its supporting infrastructure must become more efficient, better located and increasingly compatible with the needs of civilisation and the environment.

Chapter 5

The Resource Squeeze: When AI Competes with the World Around It

The previous chapter established an uncomfortable fact: artificial intelligence is physical. Every model ultimately depends on processors, electricity, cooling, buildings, networks and materials. As AI expands, the question is therefore not simply how intelligent the technology can become, but: How do we keep expanding computation without allowing digital infrastructure to compete destructively with the physical needs of civilisation? Earth is not about to run out of space for data centres, and AI is not uniquely responsible for pressure on electricity, water or land. Cities, transport, manufacturing, electrification and population growth all compete for resources. The problem is scale and concentration. If AI becomes deeply embedded in medicine, science, education, transport, robotics, government and everyday life, computing demand could continue growing for decades.

Reliability Makes the Resource Problem Harder

Data centres cannot simply consume electricity whenever it happens to be convenient. Banks, hospitals, cloud platforms and critical infrastructure require continuous availability. This means data centres need reliable grid connections, batteries, backup systems and other forms of redundancy. The challenge is therefore not merely generating enough electricity. It is delivering sufficient reliable power to the right places while homes, transport, hospitals and industries are demanding electricity from the same infrastructure. This principle will follow us beyond Earth. A future lunar or Martian data centre cannot simply install solar panels and assume the energy problem is solved. Darkness, dust, equipment failures and interruptions will require storage and backup systems. The reliability principles we develop on Earth become preparation for what comes later.

Computing Does Not Always Belong in the Same Place

One way to reduce unnecessary infrastructure pressure is to decide more intelligently **where computation should happen**. This is already happening through edge computing, which means processing information closer to where it is created instead of sending everything to a distant central data centre. An autonomous vehicle, for example, cannot send an image of a pedestrian hundreds of kilometres away and wait for another computer to decide whether it should brake. Some intelligence must remain local.

The same principle applies to factories, robots, aircraft and eventually spacecraft. A Mars rover cannot send every navigation decision to Earth and wait many minutes for instructions. It needs local computing. A future Mars data centre would take this principle much further: Mars itself becomes an edge location. Instead of Earth processing every piece of Martian information, infrastructure on Mars could process local information and continue operating when communication with Earth is delayed or unavailable.

Materials Matter Too

Computing also requires physical materials. Chips, batteries, buildings, electrical systems and cooling infrastructure depend on silicon, copper, lithium, steel, concrete and many other resources. AI therefore sits inside a much larger industrial and mining supply chain. This becomes especially important beyond Earth. Transporting every kilogram of construction material from Earth's surface would be enormously expensive. Future extraterrestrial infrastructure may therefore depend on In-Situ Resource Utilisation (ISRU)—using useful materials available at the destination. A future lunar facility, for example, might eventually use local material for shielding or construction rather than launching every structural component from Earth. The principle is simple: as civilisation moves farther from Earth, it will increasingly need to build with what is available locally.

Critical Minerals Matter Too

The material challenge extends beyond concrete, steel and silicon. Advanced technological systems also depend on a wider group of critical minerals. Batteries, power electronics, electric motors, communication equipment, renewable-energy systems and many robotic machines use materials whose supply chains can be difficult to expand quickly. Rare-earth elements are particularly important in some high-performance permanent magnets used in electric motors and generators.

AI itself does not consume rare-earth elements in the way a processor consumes electricity. The connection is indirect but important. A civilisation deploying far more AI may also deploy more data centres, electrical infrastructure, batteries, autonomous vehicles, robotic equipment, renewable-energy systems and industrial machinery. The larger this physical technological ecosystem becomes, the greater its cumulative demand for minerals and manufactured materials.

On Earth, the first response should be better mining practices, recycling, material efficiency, substitution where possible and more diversified supply chains. Extraterrestrial mining is neither necessary nor economically realistic for today's AI industry. But this book is considering technological development across centuries. If civilisation eventually operates millions of machines, power systems and computing facilities across the Solar System, continuously launching all of their raw material from Earth would make progressively less sense.

At that stage, the Moon, asteroids and eventually other accessible bodies could become additional resource regions. The objective would not necessarily be to mine space and return every valuable mineral to Earth. In many cases, the greater advantage may be using extraterrestrial materials where they are found—turning them into structures, shielding, machinery, propellant or other industrial inputs without first lifting the same mass out of Earth's gravity well.

This adds another possible progression:

Growing technological civilisation → greater material demand → more efficient terrestrial use and recycling → off-world resource utilisation → increasingly local extraterrestrial industry.

Space resources would therefore not merely support exploration. Over very long timescales, they could become part of the material supply chain supporting an increasingly computational and robotic civilisation.

Not Every Computation Is Equally Urgent

There is another way to reduce pressure: computation itself can become more flexible. Some workloads need immediate results. If a person asks an AI assistant a question, they expect a quick response. A safety system may need to react within milliseconds. But other tasks—model training, scientific processing, archival work or backups—may tolerate delay. This creates demand flexibility: shifting non-urgent computation toward times or locations where energy is more available. A future data centre might therefore operate according to principles such as:

Energy abundant → perform flexible intensive workloads.

Grid under pressure → delay non-critical computation.

Another location has surplus clean energy → move suitable workloads there.

Instead of treating electricity as an unlimited background resource, computing begins adapting itself to the energy system.

Follow the Energy

This principle could eventually extend far beyond Earth. Imagine a future network containing computing infrastructure on Earth, in orbit, around the Moon and eventually on Mars. Different locations may have different amounts of available solar or other energy at different times. Instead of always transporting energy toward computation, we could sometimes move computation toward available energy. This does not mean sending an interactive Earth user's AI request to Mars. Communication delay would make that impractical. But non-urgent scientific calculations, archival processing, backups or some forms of AI training might eventually be placed according to available energy, computing capacity and data location. Today, workload placement might mean choosing between two servers. Tomorrow it may mean choosing between data centres. One day, it could mean choosing which world.

Heat Can Become Part of the System

Data-centre heat is normally treated as waste, but in some environments it can become useful. On Earth, waste heat can sometimes contribute to heating buildings or water. Beyond Earth, integration could become even more important. A future Martian computing facility, for example, might produce heat that could potentially support nearby habitats, greenhouses, water-processing or industrial systems. Solar infrastructure powers computers; computers support robotic mining; mining supplies materials; robots maintain infrastructure; and useful waste heat supports other systems. What begins as a collection of separate technologies gradually becomes an ecosystem.

Earth Comes First

None of this means that humanity should wait for space to solve today's environmental problems. Large-scale extraterrestrial computing could take many decades or centuries to become practical, if it becomes practical at all. The first stage of the Interplanetary Cloud is therefore not interplanetary. It is better terrestrial engineering: cleaner electricity, more efficient processors and cooling, improved recycling, intelligent workload scheduling, longer-lasting hardware and responsible data-centre placement. Technological progress does not require uncontrolled consumption. Humanity cannot build an advanced space-based civilisation if it damages the civilisation that must build it. The transition therefore has to be gradual: improve computing on Earth, develop cleaner and larger energy systems, advance autonomous robotics and reusable spacecraft, establish permanent infrastructure around Earth and the Moon, learn to use extraterrestrial resources, and only then begin moving suitable computing and storage workloads beyond Earth. The sequence describes an architectural direction rather than a fixed timetable.

Saving the Present to Reach the Future

This also reconnects the resource problem with the employment problem that opened this book. If AI is used mainly to reduce labour costs, it may automate existing work while creating social disruption. But if AI also helps civilisation build new infrastructure, explore new environments and operate where humans cannot easily work, entirely new industries could emerge.

Future employment may involve designing autonomous infrastructure, planetary communications, robotic maintenance, extraterrestrial energy systems, space logistics and computing facilities operating far beyond today's terrestrial data centres. But another resource problem appears before we can reach that future. AI does not only consume electricity and materials. It consumes, produces and depends upon data. Sensors generate it. Organisations preserve it. AI systems process and create more of it. Critical information is copied for performance, backup and disaster recovery. As civilisation becomes more digital, we are not merely building more computers. We are creating an ever-growing memory of human activity—and increasingly keeping multiple copies of that memory. That is the problem we examine next.

Chapter 6

Heat: The Problem Computing Cannot Escape

If electricity is the fuel of artificial intelligence, heat is one of its unavoidable by-products. AI may appear through a cool glass screen, but somewhere physical processors performed the computation. The more computing power we concentrate into a small area, the more heat must be continuously removed.

This matters especially to the future proposed in this book. At first, space sounds like the perfect place for hot computers because space is extremely cold. But that intuition is misleading. **Space can actually make cooling difficult because there is almost no surrounding air to carry heat away.** Before imagining enormous AI data centres in orbit, on the Moon or on Mars, we therefore need to understand one fundamental rule: Computing creates heat, and that heat must go somewhere.

Why Computers Become Hot

Processors contain billions of microscopic electronic switches called **transistors**. As processors, memory, storage and networking equipment operate, they consume electrical energy, much of which ultimately becomes heat. An ordinary laptop demonstrates the principle. During light work it may remain relatively cool. Run intensive calculations and it becomes warmer while its fans work harder. A data centre experiences the same process on a vastly larger scale, with thousands of processors operating simultaneously. The problem therefore becomes not only: How much electricity can we provide? but also: How much heat can we continuously remove? More efficient processors help enormously, but they do not eliminate this problem. If each processor becomes more efficient while society installs many more processors and performs vastly more computation, total heat generation can still increase.

Cooling Moves Heat; It Does Not Destroy It

Cooling does not make heat disappear. It transfers heat from somewhere we do not want it to somewhere we can tolerate it. Heat moves principally through three mechanisms relevant to our discussion. Conduction transfers heat through physical contact. Convection uses moving liquids or gases to carry heat. Radiation releases energy as electromagnetic radiation. Earth gives engineers access to all three. Air can flow through server rooms. Liquid can carry heat directly away from powerful processors. The collected heat can then be transferred into the surrounding environment. This is why increasingly dense AI hardware is encouraging greater use of liquid cooling. Liquid can carry large amounts of heat away from processors more effectively than air alone. But even then, the fundamental problem remains unchanged:

Processor → cooling system → external environment.

The heat has merely been transported.

Space Changes the Rules

On Earth, surrounding air helps remove heat through convection. Space is almost a vacuum, so this mechanism largely disappears outside a spacecraft. This creates an apparent paradox: Space can be extremely cold and still be difficult to cool something in it. Imagine a hot cup of coffee outside on a cold day. Cold air moves around it and carries heat away. Now remove almost all of that air. The surrounding environment may be extremely cold, but you have removed one of the mechanisms that was cooling the cup. A space data centre would therefore need to collect heat from its processors, move it through cooling systems and ultimately radiate much of the unwanted heat into space.

The ISS Shows That It Can Be Done

The **International Space Station (ISS)** already demonstrates this principle. Its computers, scientific equipment, electrical systems and crew all generate heat. Because the astronauts cannot simply release that heat into surrounding air, the station uses a thermal-control system. Fluids collect and transport heat inside the station. Heat exchangers transfer it into an external cooling system, and large radiator panels ultimately release unwanted thermal energy into space as infrared radiation. This is important for the Interplanetary Cloud because it demonstrates that cooling machines in space is not science fiction. The challenge is scale. Cooling spacecraft electronics and cooling a future data centre containing enormous numbers of AI processors are very different engineering problems.

Space Data Centres Would Need Large Thermal Systems

Imagine an orbital AI data centre. Liquid cooling could collect heat from processors, pumps could transport it away from the computing modules, but the heat would still need a final destination. Large radiators—surfaces designed to release thermal energy into space—could become some of the facility's most important structures. This creates a simple relationship:

More computation → more electrical power → more heat → greater heat-rejection capacity.

A future orbital data centre might therefore look very different from a warehouse on Earth. It could consist of computing modules, solar arrays, thermal loops, radiator structures, communication systems, propulsion and robotic maintenance. In other words, thermal architecture may become just as important as computing architecture.

The Moon and Mars Create Different Problems

A lunar data centre would sit on solid ground but still have almost no useful atmosphere for convection. It would face extreme thermal conditions and could require combinations of liquid cooling, radiators, thermal storage and carefully selected locations. Lunar material could potentially provide shielding, but burying equipment would not solve the heat problem—the internally generated heat would still need a path outward.

Mars is different again. It has an atmosphere, but one far thinner than Earth's. A Martian computing facility would therefore need thermal engineering designed specifically for that environment. Mars may also provide an opportunity. Instead of treating all computing heat as waste, some could potentially support habitats, greenhouses, water processing or industrial systems. This suggests an important principle for extraterrestrial infrastructure: One system's waste can become another system's resource.

Follow the Energy—and the Ability to Cool

Earlier we considered moving flexible computing workloads toward locations where energy is abundant. Heat adds another variable. Suppose one orbital station has abundant solar electricity, but its thermal system is approaching its limit. Another has available computing capacity and greater ability to reject heat. A non-urgent workload could potentially be moved. Future workload placement might therefore consider several questions simultaneously: Where is energy available? Where is computing capacity available? Where is communication available? Where can the resulting heat be safely managed? AI itself could eventually help balance these variables. The future principle therefore becomes more precise: Do not only follow the energy. Follow the energy and the ability to cool.

Heat Is Not an Argument Against Space Computing

None of this means space-based computing is impossible. We already operate complex thermal systems in spacecraft. But scaling from today's spacecraft to enormous AI infrastructure would require much larger radiators, reliable cooling loops, robotic maintenance and technologies capable of operating for long periods far from immediate human assistance.

A credible vision of the Interplanetary Cloud therefore cannot simply say: There is plenty of space, so put the computers there. It must ask how those computers will be powered, cooled, repaired, protected and connected. The central lesson is simple: Computation creates heat wherever computation happens. Earth, orbit, the Moon and Mars cannot escape that physical reality. Our challenge is not to eliminate heat, but to collect it, move it, reuse it where practical and safely release what remains. That physical constraint will help determine not only how future data centres are built, but eventually where computing itself is allowed to happen.

Chapter 7

The Data-Centre Collision

For most of the Internet era, data centres remained largely invisible to ordinary people. We searched the web, streamed movies, stored photographs and moved businesses into the cloud without thinking much about where the computers performing those tasks were physically located. Artificial intelligence is changing that relationship. AI requires increasing amounts of computing infrastructure, while electricity, land, water and construction capacity are also required by the communities surrounding those facilities. The result is what I call the Data-Centre Collision: the point at which the digital world's demand for physical resources begins colliding visibly with the priorities of the physical world.

This does not mean data centres are inherently bad or that AI alone is responsible for pressure on infrastructure. Data centres support banking, communications, healthcare, government, science and countless other services. Electricity demand is also increasing because of population growth, manufacturing, transport electrification and other industries. The collision occurs because many demands are arriving at the same infrastructure at the same time.

Software Speed Meets Infrastructure Speed

A software application can sometimes be deployed in minutes. A new electricity transmission line cannot. This difference in timescale is fundamental. Technology companies are accustomed to software that can expand rapidly. If demand increases, additional computing resources can sometimes be provisioned almost immediately.

But the physical infrastructure underneath the cloud cannot expand at software speed. A major AI data-centre campus may require new substations, transformers, transmission capacity and additional electricity generation. All of this requires equipment, land, planning, investment and construction. The International Energy Agency estimated global data-centre electricity consumption at around 415 terawatt-hours in 2024 and projected that it could reach around 945 terawatt-hours by 2030 in its base case, with AI being the largest driver of the increase. The global number, however, hides the local problem. Data centres are concentrated in particular regions. A

country may have enough electricity overall while a particular city or grid cannot easily connect another enormous facility. AI therefore develops at software speed. Data centres develop at construction speed.
Electricity grids develop at utility speed.
Governments operate at regulatory speed.
Communities accept change at social speed. Those speeds are not the same.

When the Data Centre Becomes a Neighbor

Imagine a growing city that needs thousands of new homes, transport infrastructure and additional electricity capacity. A technology company then proposes a large AI data centre requiring a major new grid connection. The company may reasonably argue that the project brings investment, jobs and important digital infrastructure. Residents may equally reasonably ask why scarce electricity infrastructure, land or water should be committed to computing while their own community is experiencing infrastructure pressure. Neither side has to be anti-technology. They are competing for physical resources. This is the moment when the data centre stops being an invisible part of the Internet and becomes somebody's neighbour.

The Approval Problem

Large infrastructure projects also require permission. Depending on the location, a major data-centre development may require planning approval, environmental assessment, electricity connections and consultation with governments, utilities and communities. From a technology company's perspective, these processes may seem painfully slow while AI advances rapidly. From a community's perspective, they exist because a company's desire to build quickly should not automatically override the interests of people already living there. This creates a difficult tension. If approval becomes unnecessarily bureaucratic, valuable infrastructure can be delayed. But if development ignores legitimate community and environmental concerns, public resistance grows. The solution cannot simply be built faster or stop building. It has to be better planning.

Public Resistance Is Not Necessarily Anti-AI

It would be a mistake to describe every objection to a data centre as opposition to artificial intelligence. A person can support AI while opposing a particular development near their home. A community can welcome technology investment while asking a company to contribute to new electricity infrastructure. Environmental groups can recognise AI's benefits while questioning its water or energy requirements. This relates to the idea of social licence: whether the wider community continues to accept an activity and believes its benefits and costs are being shared reasonably. This matters because AI is already socially disruptive. Workers worry about employment and career stability. If the same technology is then perceived as consuming increasing quantities of electricity, water and land, resentment can grow. A dangerous public narrative could emerge: AI takes jobs, consumes resources and enriches technology companies. Whether that description is fair or not, the industry cannot afford to ignore the perception behind it.

Electricity Costs Make the Debate Personal

Climate impact can sometimes feel distant. Household electricity bills do not. It would be misleading to claim that one new data centre automatically raises everyone's electricity price. Electricity markets, generation, transmission and regulation are far more complicated than that. Large customers can also help finance new generation and infrastructure. But if communities believe they are competing with enormous data centres for electricity while household costs are rising, the political reaction can become powerful. AI infrastructure should therefore help expand the energy system rather than simply consume whatever capacity already exists. Large operators could contribute to new generation, storage and transmission, while flexible computing workloads could reduce demand during periods of grid stress. The objective should be: AI growth should help build additional capacity rather than merely compete for existing capacity.

Reliability Also Consumes Resources

Data centres require resilience. Important systems cannot disappear whenever the electricity grid fails, so facilities use batteries, generators, multiple connections and other backup systems. This reveals a broader principle: Reliability has a physical cost. A second power connection requires infrastructure. A battery requires materials. A backup generator requires equipment. A second data centre requires more land and energy. The same principle applies to information. Important data is normally replicated and backed up rather than stored in only one place. Each additional copy requires storage hardware, electricity, cooling and physical infrastructure. Later in this book, this principle will become much more important when we consider disaster recovery beyond Earth.

The Wrong Choices

One response to growing AI demand would be to build as rapidly as possible: more power generation, more transmission and more data centres, accepting whatever resource consumption follows. That could maximise computing capacity in the short term, but it could also increase environmental pressure and public resistance. The opposite response—severely restricting AI because it consumes resources—is also problematic. AI could contribute to medicine, science, climate modelling, education, engineering, energy optimisation and many other fields. It may eventually help develop the technologies needed to overcome some of the constraints discussed in this book. The choice should therefore not be: AI or the environment. It should be: How do we develop AI within environmental and social limits?

A Deliberate Pace of Evolution

AI should continue evolving, but infrastructure, society and governance need time to evolve alongside it. Maximum deployment speed does not necessarily produce maximum human benefit. A more deliberate transition gives electricity systems time to expand, processors time to become more efficient, governments time to develop sensible regulation and workers time to adapt. It also gives humanity time to develop technologies that may eventually change the physical limits of computing: advanced robotics, autonomous spacecraft, off-world manufacturing, extraterrestrial energy systems and resource extraction beyond Earth. Those technologies cannot suddenly appear when terrestrial infrastructure becomes constrained. They may require decades or longer to develop.

Earth Should Be the Launchpad, Not the Sacrifice

The argument of this book is not that Earth has run out of space for data centres. It has not. Nor is it that AI infrastructure will soon cover the planet. The concern is the long-term direction.

If AI and digital infrastructure continue expanding over centuries, Earth should not automatically be expected to absorb every stage of that expansion indefinitely. Earth supports billions of people and a biosphere we cannot presently reproduce anywhere else.

We should therefore make terrestrial computing as efficient and environmentally responsible as possible while gradually developing the ability to place appropriate future infrastructure elsewhere. Earth should be the launchpad of an expanding technological civilisation—not the resource sacrificed to sustain it. And there is another reason to think beyond today's data-centre model. Computers are not the only things multiplying. The information flowing through them is multiplying too. Humanity creates photographs, videos, scientific measurements, business records, software and sensor streams. AI adds generated content, simulations, machine telemetry and autonomous-agent records. Important information is then replicated and backed up. The next problem is therefore not simply where we will put more computers. It is where we will put civilisation's expanding memory—and how many copies of that memory we will need to keep.

Chapter 8

When Compute Competes with Civilisation

The previous chapter described the **Data-Centre Collision**: the point at which digital infrastructure begins visibly competing with communities for electricity, land, water and physical infrastructure. But there is a larger question behind that collision. What happens if computing demand continues growing not merely for the next five or ten years, but for generations? This book does not assume that today's AI growth will continue indefinitely. Technology becomes more efficient, markets change, regulation develops and long-term predictions are often wrong. Nor is humanity about to run out of physical land for computers. The concern is different. Civilisation has many competing uses for finite infrastructure and resources. If computing becomes one of humanity's major industrial activities, societies will increasingly have to decide how much electricity, manufacturing capacity, water, land and material should support machines—and how much should remain available for everything else civilisation needs. That is when compute begins competing with civilisation.

Billions of People—and Far More Machines

Current United Nations projections expect the world's population to peak at around **10.3 billion during the mid-2080s** before declining slightly by 2100. But the number of intelligent machines does not have to follow human population. One person may interact with many AI systems. Vehicles may contain AI. Factories may operate robotic fleets. Hospitals may use diagnostic systems. Businesses may operate thousands of software agents, while cities increasingly depend on intelligent infrastructure. Not every intelligent system requires its own data centre, of course. Computing resources are shared. But collectively these systems require processing, storage, communication and energy. The long-term question therefore becomes: How much computation will civilisation eventually require? We do not know. That uncertainty is precisely why thinking about the architecture before serious constraints emerge is worthwhile.

Human Needs Come First

One principle should sit underneath everything proposed in this book: Computing infrastructure exists to serve civilisation. Civilisation does not exist to serve computing infrastructure.

If a region does not have sufficient electricity for homes, hospitals and essential services, a major AI facility should not simply consume the remaining capacity without contributing to additional infrastructure. If water is scarce, cooling systems should reflect that reality. If strategically located land is urgently needed for housing or essential infrastructure, those needs must be considered.

This does not mean AI always loses whenever resources are contested. AI itself can support medicine, science, energy systems, governments and businesses. The question is whether the value created by computing justifies the resources required to provide it.

The Invisible Cost of Redundancy

Computing demand also grows for a reason that ordinary users rarely see: Reliable systems deliberately duplicate things. A company with 100 terabytes of important information is unlikely to store only 100 terabytes physically. It may maintain multiple local copies, another copy in a separate data centre and additional backups elsewhere. That is not necessarily waste. It is resilience. But resilience consumes resources. The same principle applies to computing infrastructure itself. Important applications may run on multiple servers. Data centres may have redundant power, cooling and network systems. This creates a paradox: The more dependent civilisation becomes on data, the more copies of that data civilisation may need to protect.

Civilisation Is Becoming a Recording Machine

At the same time, humanity is recording more of what it does. A modern railway can generate signalling information, CCTV footage, maintenance records, operational logs and sensor telemetry. Vehicles, factories, hospitals, satellites, buildings and robots generate their own streams of information. AI adds another layer. AI systems create text, images, software, simulations and operational records. Autonomous agents may also produce extensive audit trails explaining what they observed and what actions they took. Civilisation is therefore not merely doing more. It is recording more of what it does.

If intelligent machines become embedded throughout society, this digital history could expand enormously.

We Also Need to Learn How to Forget

The answer cannot simply be infinitely larger storage systems. Much of the information generated by machines has little permanent value. A sensor recording temperature every second may not need every individual reading preserved forever. Temporary AI calculations may become irrelevant once a result has been verified. Future infrastructure therefore needs better storage—but it also needs better forgetting. AI itself could help classify information according to importance, legal requirements, scientific value and likely future usefulness. Some information remains immediately available. Some is compressed. Some moves into long-term archives. Some is safely deleted. The most sustainable byte may sometimes be the byte we correctly determine we no longer need to store.

Not Every Computation Is Equally Urgent

The same principle applies to processing. Imagine an electricity grid under severe pressure during an extreme heatwave. Hospitals require electricity. Homes require cooling. Public transport needs power. At the same time, data centres may be running emergency services, scientific workloads, model training and entertainment applications. From a processor's perspective, they are all calculations. From civilisation's perspective, they may not be equally urgent. Future computing systems could therefore become more flexible about when certain workloads run. Emergency and safety-critical systems continue immediately. Non-urgent training or scientific processing may wait until electricity is more abundant. The objective is not to ration intelligence. It is to make computing a responsible participant in the infrastructure system surrounding it.

The Planet Becomes Part of the Architecture

This leads to a broader concept that we can call **planet-aware computing**. Traditional computing architecture asks: Where should the application run? Where should the database be stored? What happens if a server fails? Future sustainable computing may also ask: Where is clean energy available? Which electricity grid is under pressure? Where is water scarce? Which region is experiencing extreme heat? Where can waste heat be reused? Which workloads can wait? Where is the required data already stored? The physical condition of the planet therefore becomes an input into computing decisions. This should happen long before humanity considers moving significant computing infrastructure into space.

Climate Change Changes the Risk Map

There is another reason geography matters. Data centres may operate for decades, while the environments surrounding them can change. Extreme heat can increase cooling demand. Floods, wildfires and storms can threaten buildings, electricity networks and communications. Sea-level rise may alter the long-term risk of some coastal infrastructure.

Climate change does not mean that every disaster is caused by climate change, nor does every hazard respond to it in the same way. But long-lived infrastructure cannot assume that tomorrow's risk will always look like yesterday's. This strengthens the existing engineering principle of geographic redundancy. Important digital information should not depend entirely on one vulnerable location. Today that means another data centre, city or region. Over much longer timescales, humanity may eventually ask a much larger question: Should every irreplaceable copy of civilisation's knowledge remain on the same planet?

From Geographic Resilience to a Bigger Map

Today, extraterrestrial backup would be extraordinarily expensive and unnecessary for ordinary systems. Terrestrial regions provide vastly more practical disaster recovery. But this book is considering infrastructure across centuries. A future civilisation far more dependent on digital knowledge might eventually develop a hierarchy such as:

Local copy → Regional copy → Continental copy → Global archive → Extraterrestrial archive

The farther away a copy is, the less current it may need to be. A lunar archive, for example, would not replace normal backups. It could eventually become another layer protecting particularly valuable scientific, cultural or technological information against events confined to Earth. Exactly how information could move reliably between Earth, orbital infrastructure, the Moon and eventually Mars is a much larger engineering problem. We will return to that later when we examine interplanetary communications, moving relay networks and delay-tolerant networking. For now, the important idea is simpler: As civilisation becomes more dependent on digital information, resilience may eventually need a larger geographic map.

Civilisation Needs a Bigger Map

Better processors should come first. Cleaner energy should come first. Efficient cooling, responsible data retention, flexible workloads and better terrestrial planning should come first. Space is not an excuse to avoid any of them. But if humanity continues developing AI and digital infrastructure over centuries, the architecture itself may eventually expand. Earth could remain the centre of human civilisation while the Moon, orbital infrastructure and eventually other planets become extensions of its computing, storage and communication systems.

The purpose would not be to abandon Earth. It would be to prevent Earth from eventually carrying every physical consequence of technological civilisation by itself. And this reconnects the argument with the employment disruption that opened this book. The engineers concerned today that AI can generate their code may eventually be needed to design computing systems spanning enormous distances. Network engineers may confront moving planetary endpoints. Cooling engineers may design thermal systems for orbital computing. Electricians and energy engineers may work on infrastructure far beyond terrestrial data centres.

AI will automate parts of these professions. But technological expansion also creates new engineering problems. New problems create new work. The challenge is reaching that future with a healthy planet, functioning societies and enough time for people and infrastructure to adapt. Saving the present and building the future are therefore not separate projects. They are the same project at different timescales.

PART III SAVING THE PRESENT

A distant future is useless if the present is mismanaged

Closing Part II

Part II showed that computation has physical consequences. More AI can mean more electricity, heat, infrastructure and competition for resources, even as efficiency improves. Earth therefore cannot be treated as an unlimited substrate for digital expansion. Yet the answer is not simply to restrict technological progress. Before imagining infrastructure beyond Earth, we first have to protect the civilisation capable of building it. Part III turns from identifying the problem to managing the transition.

What PART III Covers

A distant technological future has little value if humanity damages the environmental, economic and social foundations required to reach it. This Part therefore puts Earth first. It examines responsible AI expansion, the limits of deploying enormous autonomous machines on a populated planet, and why space may eventually offer an industrial frontier better suited to some large-scale robotic activity. It also develops the Employment Bridge: the idea that AI should not merely reduce the number of people required to perform today's work, but help humanity create new industries, infrastructure and objectives. The goal is responsible acceleration—protecting people and the planet today while developing the capabilities that could expand human opportunity tomorrow.

Chapter 9

AI Cannot Flourish on a Damaged Planet

It is tempting to discuss the future of artificial intelligence as though technological progress were guaranteed. Processors will become faster. AI will become more capable. Robots will become more autonomous. Rockets will become cheaper. Humanity will establish infrastructure on the Moon and perhaps eventually Mars.

But none of this is guaranteed. Every advanced technology depends on something more fundamental: a sufficiently stable civilisation capable of building it. AI requires electricity grids, semiconductor factories, communication networks, universities, engineers, manufacturing supply chains, investment and functioning institutions. Space exploration requires even more. A civilisation repeatedly struggling with environmental disasters, damaged infrastructure, conflict, severe resource shortages and economic instability will have fewer resources available for ambitious technological projects. This leads to one of the central arguments of this book:

If we want AI to flourish in the future, we must protect the civilisation and planet on which AI is being created today.

AI and Climate Change Are Different Problems

Artificial intelligence did not create climate change. The principal cause of modern climate change is the accumulation of greenhouse gases from human activities, particularly the burning of fossil fuels. The IPCC has concluded that human activities have unequivocally caused global warming. AI becomes connected to this much larger problem because computation requires electricity. If additional electricity comes from high-emission sources, expanding computation can contribute indirectly to greenhouse-gas emissions. If it comes from low-carbon sources, that impact can be substantially reduced. The relationship is therefore:

More AI → More computation → More electricity demand → Environmental impact depends strongly on how that electricity is produced.

AI's environmental impact cannot be separated from the infrastructure supporting it.

Saving the Present Does Not Mean Stopping AI

The answer is not simply to stop developing artificial intelligence. AI itself may help address some of the problems confronting civilisation. It can contribute to scientific modelling, electricity-grid optimisation, materials research, industrial efficiency, engineering and analysis of enormous datasets. The question should therefore not be: AI or the environment? It should be: How do we develop AI without allowing its physical infrastructure to undermine the environment and society that support it? Efficiency should come before unnecessary expansion. Cleaner electricity should accompany increasing computational demand. Hardware should be used responsibly and kept useful for as long as practical. Flexible workloads should adapt where possible to available energy. These ideas have already been explored in earlier chapters. The larger principle is simple: AI infrastructure must learn to adapt to the planet rather than expecting the planet to adapt endlessly to AI.

Space Is Not an Excuse

This distinction becomes especially important because this book will eventually propose infrastructure beyond Earth.

If humanity someday builds data centres on the Moon, that does not give us permission to waste electricity on Earth. If robots eventually extract extraterrestrial resources, that does not justify irresponsible resource use here. If we someday generate large quantities of energy in space, that does not remove the need to improve terrestrial energy systems.

And Mars is not a convenient replacement for Earth. Earth provides breathable air, liquid water, a protective atmosphere, a magnetic environment, ecosystems and conditions in which billions of humans can live without spacesuits. Our future expansion should protect that extraordinary advantage rather than treat Earth as something civilisation can simply replace. Space should extend Earth's technological capacity, not excuse damage to Earth.

AI Needs Humanity to Want It

There is another requirement for AI's future that receives less attention than processors, electricity or algorithms: AI needs social legitimacy. People must believe that the technology creates enough benefit to justify its costs. A worker whose job changes because of AI needs a realistic opportunity to move toward another useful role. A community hosting major computing infrastructure should experience benefits rather than only resource pressure. Environmental impacts should be measured honestly. Electricity infrastructure should expand alongside major new demand.

If society increasingly sees AI as a technology that removes jobs while consuming electricity, water and land for the benefit of a relatively small number of companies, political resistance will grow.

If people instead see AI improving services, advancing science, helping them work, creating industries and solving difficult problems while its environmental footprint is responsibly managed, its relationship with society can be very different.

The future of AI is therefore not only a computer-science problem. It is also a social-design problem—not manipulating society into accepting AI but designing AI development so that society genuinely benefits from it.

Saving Jobs Does Not Mean Freezing Jobs

The same principle applies to employment. Trying to preserve every occupation exactly as it exists today is unrealistic. Technology has always changed work. But allowing change to happen faster than people can reasonably adapt can create unnecessary economic and social damage. The goal should therefore be transition rather than preservation. A DevOps engineer may increasingly supervise AI-generated infrastructure instead of manually writing every configuration. A software engineer may spend more time designing and validating systems while AI generates routine code. Other professions will experience their own versions of this transition.

Some roles will shrink. Others will expand. New ones will appear. The difficult part is the period between those states. Education systems, employers, governments and workers therefore need time to adapt. AI development should not be judged only by how rapidly a task can be automated, but also by whether society can productively absorb the resulting change.

The Problems AI Creates Can Also Create Work

Technology has repeatedly created new problems alongside new capabilities. The Internet created enormous cybersecurity challenges—and cybersecurity became an industry. Cloud computing created new complexity around distributed systems—and professions such as cloud engineering, DevOps, Site Reliability Engineering and platform engineering expanded. AI will create its own problems involving security, governance, energy, verification, automation, robotics and infrastructure. Those problems require solutions. And if humanity eventually expands AI infrastructure beyond Earth, the engineering problems become much larger. Someone—or some combination of humans and AI—will need to design lunar energy systems, coordinate robotic construction, operate remote data centres, manage interplanetary communications, engineer thermal systems and secure autonomous machines operating hundreds of thousands or millions of kilometres away. AI may automate parts of these professions too. But expansion continually creates problems that did not previously exist. New technological frontiers create new work because they create new problems to solve.

AI and Humans Need a Larger Project

Perhaps this provides another way to think about today's employment anxiety.

Much of AI is currently being inserted into economic structures that already exist. We have software teams, so AI generates code. We have accounting departments, so AI automates accounting tasks. We have customer-service centres, so AI answers customer questions. AI enters an existing economic box and performs some of the work already inside it. Human and machine therefore appear to be competing for the same activity. But what happens if technology helps us make the box larger? A human cannot spend months continuously operating machinery on an airless lunar surface. Robots potentially can. A person on Earth cannot directly control a Mars robot in real time because communication takes minutes. Autonomous systems can make some decisions locally. A human cannot manually monitor millions of components spread across an interplanetary infrastructure. AI could. In those environments, AI becomes less about replacing a person performing an existing task and more about enabling activities that humans could not realistically perform alone. That distinction is central to the future explored in this book.

A Controlled Transition, Not a Race Without Limits

AI development is often described as a race: companies compete for more capable models, countries compete for technological leadership, and manufacturers compete to produce more advanced processors. Competition can accelerate innovation. But civilisation also needs coordination. Electricity grids need time to expand. Cleaner energy needs time to develop. Workers need time to adapt. Communities need a voice in major infrastructure decisions. Governments need time to establish rules without freezing useful innovation. A society experiencing job disruption, rising infrastructure pressure and environmental anxiety simultaneously may become increasingly hostile toward AI. That would be bad for people. It would also be bad for AI. A technology cannot flourish indefinitely without sufficient social acceptance.

First, We Have to Earn That Future

The long-term vision of this book is ambitious: computing and AI infrastructure extending from Earth into orbit, the Moon, Mars and perhaps eventually farther into the Solar System. But we are not there yet. The foundation is being built now. We need cleaner energy, efficient computing, responsible resource use, resilient infrastructure, workers capable of adapting to automation and governance that protects society without preventing beneficial innovation.

If we damage that foundation badly enough, the future built upon it becomes harder to reach.

That is why **saving the present and flourishing with AI are not competing goals**. Saving the present is what makes the future possible. And one of the most important systems determining whether we can achieve that balance is already under pressure today: the electricity grid. The grid was built to power an industrial and increasingly digital civilisation. Now it must evolve to support an increasingly computational civilisation as well.

Chapter 10

Earth vs Huge Autonomous Machines

The previous chapter argued that AI cannot flourish on a damaged planet. That does not mean humanity should respond by stopping technological progress. Artificial intelligence could become one of the most useful technologies we have ever developed. It may accelerate science, medicine, engineering and productivity and eventually help humanity operate in environments where people cannot safely work alone.

The problem is not acceleration itself. The problem is acceleration without preparation. I call the alternative Responsible Acceleration: continuing to advance AI while deliberately developing the energy, infrastructure, workforce, security and governance systems required to support it. The objective is neither to race ahead regardless of consequences nor to freeze technological development out of fear. Our ability to manage AI needs to develop alongside our ability to build it.

The Adaptation Gap

AI can change much faster than the physical and social systems surrounding it. A new model can reach millions of users through software. A new power station or transmission line may take years to build. Developing an experienced engineer takes years. Education systems, laws and workplace structures also take time to change. This creates what we can call the adaptation gap: the distance between the speed of technological change and the speed at which society can respond.

If AI capabilities accelerate while electricity infrastructure, workforce skills, cybersecurity and regulation develop much more slowly, problems accumulate inside that gap. Companies may automate work before employees have realistic pathways into new roles. Data centres may demand electricity faster than grids can expand. AI agents may receive operational authority before organisations develop appropriate controls.

Responsible acceleration means continuing technological progress while trying to prevent this gap from becoming dangerously large.

Experiment Fast, Deploy Carefully

Not every use of AI carries the same consequences. An AI summarising a document presents a very different risk from an AI controlling railway infrastructure or a large industrial robot. The appropriate speed of deployment should therefore depend partly on what happens if the system is wrong. This leads to a simple principle: Experiment fast. Deploy carefully. Researchers and engineers need freedom to experiment. A controlled sandbox allows an AI system to fail, misunderstand instructions and attempt incorrect actions without affecting important real-world systems. Deployment is different. Once AI can affect people, money, infrastructure or physical safety, stronger testing and supervision become necessary. Low-risk systems can progress rapidly. Higher-risk systems require stronger verification. Safety-critical systems may require extensive testing and human oversight. This is not opposition to innovation. It is how mature engineering handles consequences.

People Need Transition Time Too

Responsible acceleration cannot concern only machines and infrastructure. People also need time to adapt. Consider a software or infrastructure engineer whose employer introduces increasingly capable AI coding agents. If management immediately concludes that the same work can now be performed with fewer people, the short-term financial calculation may appear attractive. But other questions follow. Who understands the system when AI produces an incorrect solution? Where do future senior engineers come from if junior positions disappear? Who takes responsibility for AI-generated code entering critical infrastructure? A better transition begins by using AI to increase human capability. An engineer may move from manually producing every configuration toward designing systems, reviewing AI-generated work, testing it, securing it and understanding its operational consequences. The role changes rather than simply disappearing.

My Own Profession Illustrates the Change

I can already see this transition in the kind of work I have done myself. As a platform and infrastructure engineer, I have spent years working with cloud infrastructure, networking, automation, DevOps, Site Reliability Engineering, Kubernetes and Terraform. AI can now generate portions of configuration, analyse errors and suggest solutions in seconds.

Refusing to use those tools because they perform tasks that once required more manual effort would make little sense. But blindly copying their output would make equally little sense. The more useful path is to move upward in the problem. AI can help produce an initial configuration while the engineer concentrates on architecture, reliability, security, cost and operational consequences. It can accelerate troubleshooting while the engineer retains enough understanding to recognise when its conclusion is wrong. The important skill therefore begins moving from: "Can I produce this configuration?" toward: "Can I design, verify and operate the complete system this configuration belongs to?" AI may still reduce demand for some roles. Responsible acceleration cannot guarantee that every job survives. But it creates another possibility: use increased productivity to attempt work that organisations and civilisation previously lacked the capacity to perform.

Productivity Can Reduce Work—or Expand Ambition

Imagine ten people produce 100 units of useful work. AI makes them substantially more productive. An organisation can respond: We can produce the same output with fewer people. Or it can ask: What could these people build now that their productive capacity has increased? Real organisations will do both. Cost reduction is sometimes necessary. But at civilisation scale, the distinction matters enormously.

If most AI productivity gains are converted only into reducing human labour, AI can intensify employment disruption and social resistance. If some of those gains allow humanity to tackle larger problems, new industries and professions can emerge.

Those larger projects do not need to begin in space. They can include cleaner electricity systems, climate adaptation, scientific research, advanced medicine, resilient infrastructure and environmental restoration. Over much longer timescales, the projects can become far more ambitious. AI should not only help us perform today's work with fewer people. It should help humanity attempt work that was previously beyond our practical capacity.

Infrastructure Must Accelerate Too

A more technological civilisation will require substantial energy and physical infrastructure. AI is only one part of that demand. The answer therefore cannot simply be to consume less energy forever. Humanity also needs to become better at producing large quantities of energy with progressively lower environmental impact. The same principle applies to computing infrastructure.

If a large data centre creates substantial new electricity demand, the infrastructure supporting that demand should expand with it. Computing growth and energy growth increasingly need to be planned together.

Responsible acceleration therefore combines two ideas: Use resources more efficiently while expanding cleaner capacity. Efficiency gives civilisation breathing room. Greater clean-energy capacity gives civilisation room to grow.

Governance Should Follow Consequence and Authority

AI also needs governance, but regulation should concentrate heavily on what systems are actually permitted to do. Important questions include: What information can the AI access? What actions can it perform? How are consequential decisions recorded? Can humans override it? Who is responsible when something goes wrong? The important question is therefore not merely: How intelligent is the AI? It is also: What authority does the AI have over the real world? A highly capable AI isolated inside a research environment presents a different immediate risk from a less capable system authorised to modify critical infrastructure. Capability matters. Authority matters too.

Autonomy Should Be Earned

This leads to another principle: AI should earn autonomy through demonstrated reliability. A new agent might begin by making recommendations. Humans review them. If its performance becomes reliable, it may receive permission to perform low-risk actions. Its actions remain logged and failures are studied. Greater authority is granted only where evidence justifies it. This resembles how organisations treat people. A new employee is not normally given unrestricted administrative access to every system on the first day. Responsibility expands with role, experience and demonstrated reliability. AI should be treated with similar discipline. This becomes even more important when AI controls machines. An incorrect paragraph is one kind of failure. An incorrect decision by a twenty-tonne autonomous machine is another.

Space as an Industrial Sandbox

There is another reason advanced robotics may eventually develop differently beyond Earth. On Earth, technological development operates inside necessary limits. Nearly ten billion people may eventually share this planet, together with ecosystems, agriculture, cities, transport networks and other infrastructure. A failure involving a small experimental robot can usually be contained. A failure involving a highly autonomous industrial machine weighing hundreds or thousands of tonnes could be very different. Governments, communities and environmental organisations would reasonably ask what happens if such a machine loses control, damages infrastructure, contaminates an environment or operates in a way its designers did not anticipate.

These constraints are not obstacles that should simply be removed. Many exist because Earth is the only habitable planet humanity currently possesses. Protecting people and the biosphere should take priority over testing increasingly powerful machines simply because engineers are capable of building them. This may place practical limits on how quickly fully autonomous, very large robotic industrial systems can be deployed near populated or environmentally sensitive areas.

Selected extraterrestrial environments could eventually provide a different kind of engineering sandbox. A remote lunar mining region, for example, could potentially contain enormous excavators, autonomous transport systems, robotic construction equipment, processing plants and machine fleets operating across areas with no nearby city, agricultural land or terrestrial ecosystem. Mars and some asteroids could eventually provide other environments for specialised robotic industry. Regulations, safety controls and responsible governance would still be necessary, but the consequences being protected against would be different.

This could allow robotics to progress beyond individual machines toward something much larger: **autonomous industrial ecosystems**. Mining robots obtain material. Transport machines move it. Processing systems refine it. Manufacturing equipment produces structures and components. Construction robots assemble new facilities. Inspection robots identify failures. Maintenance machines perform repairs, while local AI coordinates the overall system. Instead of building one intelligent robot, civilisation begins learning how to operate networks of machines that collectively maintain physical industry.

Earth = living environment that must be protected.
Space = selected controlled industrial environments where autonomy can eventually be tested at much greater physical scale.

Then move from individual autonomous machines to:

Autonomous machine → Robot fleet → Autonomous industrial ecosystem

Autonomous Does Not Mean Human-Free

That future should not be confused with removing humans completely. In fact, the earliest large extraterrestrial industrial systems may require substantial human involvement. Robots will fail in ways their designers did not predict. A machine may become physically trapped, a component may break in an unusual way, several failures may interact, or a repair may require improvisation beyond the capability of available robots.

From Earth, humans can supervise many operations, but distance eventually becomes a serious limitation. Lunar infrastructure is close enough for substantial Earth-based support despite communication delay. Mars is very different: communication takes minutes each way depending on planetary positions, making real-time remote control impossible.

Early major extraterrestrial industrial regions could therefore combine autonomy with human presence. Robots perform most repetitive, dangerous and physically demanding work, while humans supervise operations, investigate unusual failures, perform repairs beyond robotic capability, install upgrades and make decisions that exceed the authority of local AI.

Permanent or semi-permanent settlements could eventually develop around some of these facilities. Teams might work in rotations, much as people today rotate through remote research stations, offshore installations or other difficult environments. Over time, greater robotic capability could reduce the number of humans required for routine operation, but it would not necessarily eliminate human presence. Instead, the relationship changes: **humans manage exceptions while machines perform increasingly large amounts of normal physical work.**

This creates another developmental path:

Human-operated machines → remotely supervised robots → autonomous robot fleets → human-supervised industrial ecosystems → increasingly self-maintaining extraterrestrial infrastructure

The importance of this progression extends beyond mining. The same industrial ecosystem could eventually construct solar arrays, communication stations, habitats, launch facilities and data centres. A large lunar data centre would therefore not appear as an isolated building delivered from Earth. It could become one output of a much larger robotic industrial system already capable of extracting material, generating energy, transporting equipment, constructing structures and maintaining machinery.

In that sense, extraterrestrial space offers more than additional land. It provides environments in which humanity may eventually learn to operate physical systems at scales and levels of autonomy that would be much more difficult to develop responsibly on a crowded living planet. Earth remains where we protect life and develop the foundations. Selected environments beyond Earth become places where some of those technologies can eventually expand to much greater physical scale.

Responsible Acceleration Is a Sequence

The future proposed in this book cannot happen in one enormous leap. It is a progression:

AI assists humans → AI agents perform tasks → agents coordinate with agents → robots gain greater local autonomy → autonomous systems operate remote infrastructure → space infrastructure expands → local resources support larger facilities → computing infrastructure follows.

Every stage depends on knowledge gained from the previous one. That is Responsible Acceleration at civilisation scale.

The Goal Is Not Less AI

The central argument of this chapter can therefore be summarised simply: We do not necessarily need less AI. We need the systems supporting AI to evolve with it. More intelligence requires efficient computing. More computing requires appropriate energy. More automation requires better security. More autonomy requires stronger verification. More productivity requires new economic opportunities. More robotic capability requires safer experimentation. And greater technological power requires greater responsibility.

If we achieve that balance, the AI transition does not have to become a competition in which humans lose so machines can advance.

AI can instead help civilisation attempt larger things. But building those larger things beyond Earth requires one resource above almost everything else: energy. The Sun provides vastly more energy than civilisation currently consumes, and beyond Earth sunlight could eventually become one of the principal energy sources for robotic infrastructure, communications and computing. Yet using solar energy beyond Earth is not as simple as placing a solar panel in sunlight. Planets rotate. Moons move. Spacecraft enter shadow. Mars experiences dust. Distance from the Sun matters. A future data centre cannot simply shut down whenever its solar panels become dark. The next challenge is therefore to understand how an increasingly computational civilisation could capture, store and eventually redirect solar energy across environments far beyond today's terrestrial data centres.

Chapter 11

Space as the Industrial Frontier

Why Not Move Billions of People into Space Instead?

Science fiction often imagines a different solution to Earth's future pressures. Instead of moving industry away from Earth, why not move people?

Films and television have imagined enormous orbital habitats, permanent cities on Mars and spacecraft carrying generations of people between worlds. Such ideas are useful because they encourage us to think beyond today's limits. But building a place where machines can operate is very different from building another Earth.

A data centre does not need forests, oceans, rainfall or breathable air. A mining robot does not need a blue sky. An autonomous factory does not suffer from isolation. Humans do.

Earth provides an extraordinary life-support system without requiring us to manufacture one. We can walk outside and breathe. Plants grow in open environments. Water moves through natural cycles. Vast ecosystems support agriculture and biodiversity. People can travel between different climates, landscapes, cities and countries.

On Mars, almost every part of that environment would have to be engineered. Humans would require protected habitats, breathable air, temperature control, water systems, food production, radiation protection, medical facilities and reliable energy. Long-duration isolation and confinement could also create psychological and social challenges. A large orbital settlement would face many of the same problems.

This does not mean permanent human settlements beyond Earth are impossible. Future technologies may make them increasingly practical, and some people may eventually choose to live away from Earth for very long periods. But this book does not assume that billions of people must leave Earth or that Mars must become a replacement planet.

There is another possibility.

Instead of moving the population away from Earth's natural environment, we can gradually move more of the infrastructure that does not require that environment.

Large data centres, specialised manufacturing, resource processing, energy infrastructure and eventually some forms of heavy autonomous industry could increasingly operate beyond Earth where the engineering and economics justify it. Humans could still travel to orbital facilities, the Moon and Mars for scientific, engineering and operational assignments, while many positions operate through rotating crews rather than requiring the same individuals to remain there permanently.

In this model, **the infrastructure can be permanent even when the people are not**.

Earth remains home. Space becomes an extension of civilisation's industrial and technological capacity.

This distinction may become increasingly important as humanity confronts population, food, energy, housing and environmental pressures. Global population growth is expected to slow over the long term, although the exact peak and future trajectory remain uncertain. Even without indefinite population growth, supporting billions of people at a high standard of living while protecting natural ecosystems will remain a major challenge.

Land on Earth therefore has value far beyond the buildings we can place upon it. Land can produce food. It can support forests, wetlands and biodiversity. It can provide space for cities and recreation. It can also be restored after previous industrial use.

This creates a different long-term objective:

Do not abandon Earth because civilisation becomes more technologically advanced. Use technology to reduce the physical burden civilisation places upon Earth.

That transition would take generations. In the nearer term, data centres and industry will overwhelmingly remain terrestrial. More efficient computing, cleaner energy, recycling, better urban design and environmental protection remain essential. Orbital computing and lunar infrastructure could develop gradually. Human missions and settlements could expand alongside them. Mars would come later and would be vastly more difficult.

Over decades and potentially centuries, however, the balance could begin to change. Some activities that once had to occur on Earth may no longer need to occupy Earth's land or consume exclusively terrestrial resources.

Humans would still travel through this wider civilisation. Engineers might spend years at a lunar facility before returning home. Specialists could eventually undertake long Martian assignments before another team takes over. Scientists could work aboard orbital stations. As transportation improves, people might move repeatedly between these environments.

But this is very different from assuming that humanity's future requires billions of people to spend their entire lives inside artificial habitats.

The Solar System contains enormous space and resources. What it does not obviously contain nearby is another Earth.

For that reason, one of the priorities of an interplanetary civilisation may be surprisingly conservative:

Protect the planet that already supports human life naturally, while gradually moving appropriate machines, industry, computing and resource demand into environments where humans do not need to live permanently.

Chapter 12

The Employment Bridge

Artificial intelligence may eventually create industries and occupations that barely exist today. But that does not solve the problem facing workers whose jobs are changing now. This is one of the hardest parts of technological transition. It is easy to say that previous technological revolutions eventually created new work. That may be true over long periods, but it provides little comfort to someone whose position disappears today. The replacement job may require different skills, exist somewhere else, pay differently or not yet exist at all.

A worker cannot simply jump from an occupation being disrupted today into an industry that may take decades to develop. There has to be something between them.

I call this the **Employment Bridge**: the pathway from work being changed by AI today toward the much larger range of work that humans, AI and robots could perform together tomorrow.

But building that bridge requires changing one assumption. The purpose of AI should not simply be to accomplish today's work with fewer people. It should also help civilisation accomplish far more work than it could afford to attempt before.

The Problem Is Not Only Automation

The employment question is often reduced to whether AI will destroy more jobs than it creates. Nobody can answer that reliably because we do not know what industries, technologies or occupations will exist several decades from now.

The more immediate problem is timing.

AI software can be deployed quickly. A company may automate part of a workflow within months, while building new industries, electricity networks, manufacturing systems, robotics sectors or space infrastructure takes years or decades. Jobs can therefore change faster than replacement opportunities appear.

There is also a powerful economic incentive behind this imbalance. If AI allows a company to produce the same output with fewer employees, reducing labour costs can produce an immediate financial benefit. Investing in a completely new industry may take years before producing a return.

That does not mean companies are wrong to pursue productivity. Productivity is one of the foundations of economic progress. The larger question is what we do with the productivity AI creates.

Imagine an organisation in which AI allows ten people to perform work that previously required fifteen. One response is:

Same objective → greater productivity → fewer workers

Sometimes that will be economically unavoidable.

But there is another possibility:

Greater productivity → larger objectives → new products and infrastructure → new work

If technology makes civilisation more productive, we do not necessarily have to keep our ambitions fixed. Instead of asking only how many people AI can replace, we can ask what problems humans can now afford to solve because AI and robotics have made us more capable.

If most productivity gains are used only to reduce labour requirements for existing activities, AI may become associated primarily with displacement. If some of those gains are reinvested into solving problems that previously remained neglected, productivity can expand the amount of useful work civilisation attempts.

There is no shortage of such work.

Give the Worker a Robot, Not Just a Redundancy Notice

Consider sanitation. In parts of the world, people still perform dirty, dangerous and physically exhausting work cleaning drains, sewers and waste systems. Technology should not simply ask how the worker can be removed from the payroll. It should ask how the worker can be removed from the **danger**.

Imagine a sanitation worker who previously relied heavily on manual tools. Instead, that worker supervises several specialised machines. A small inspection robot enters a drain and identifies a blockage. Another machine breaks apart or extracts accumulated waste. A robotic system moves the material toward a collection point, while automated or semi-automated machinery transfers it for treatment or disposal.

The worker has not become useless. The nature of the work has changed:

Manual cleaner → equipment operator → robot supervisor → sanitation infrastructure specialist

This is an important distinction between automating a worker and automating the dangerous part of the worker's job.

The same principle could apply to polluted rivers and waterways. Robotic systems could inspect contaminated areas, collect floating waste, monitor water quality and operate in locations that are unpleasant or unsafe for people. Humans would still determine objectives, supervise operations, maintain equipment, investigate unusual conditions and make decisions when automated systems encounter situations they do not understand.

Clean water creates another enormous field of useful work. Sensors and AI could monitor treatment systems, identify leaks, detect unusual contamination patterns and help coordinate maintenance. Robots could inspect pipes, drains, reservoirs and other infrastructure.

The objective is not to create artificial employment. **The problems already exist. Technology gives us new ways to solve them.**

Every Robot Creates an Industry Around It

A sanitation robot does not appear because someone asks an AI model to clean a drain.

Someone must design the machine. Someone must manufacture its motors, sensors, cameras, batteries and structural components. Software engineers must build its control systems. AI specialists must develop perception and decision-making systems. Network engineers must connect fleets of machines. Cybersecurity specialists must prevent them from being misused. Mechanical and electrical technicians must maintain them. People must manufacture replacement parts, manage logistics, train operators, analyse failures and improve the next generation.

A large deployment of environmental robots therefore creates an ecosystem:

Problem → robot → software → manufacturing → energy → communications → maintenance → training → new employment

Multiply that across sanitation, water, agriculture, environmental restoration, infrastructure inspection, disaster response, construction, energy systems and advanced manufacturing, and the Employment Bridge becomes much larger than simply teaching displaced office workers how to use another piece of software.

We begin creating new physical industries around intelligence.

Build the Bridge Before the Job Disappears

This transition should begin while people are still employed. Waiting until someone's position has disappeared and then telling that person to retrain places most of the burden on the worker.

If an organisation knows that AI is likely to change a department substantially over the next several years, preparation should begin before the disruption is complete.

Some employees may learn to supervise AI-supported systems. Others may move toward security, governance, customer interaction, robotics, infrastructure or specialist roles. Some will move into entirely different industries.

The objective should not be to turn everyone into an AI engineer.

A railway worker understands railway operations. An accountant understands financial processes. An electrician understands electrical systems. A nurse understands healthcare environments. A sanitation worker understands drainage and waste problems that a robotics engineer sitting in a laboratory may never have encountered.

That knowledge has value.

The stronger combination is often:

Existing domain knowledge + new technological capability

The people who understand the problem should help shape the machines designed to solve it.

AI Literacy Will Become Ordinary

AI literacy may eventually become similar to computer literacy. Several decades ago, operating computers was a specialised capability. Eventually computers became ordinary tools across almost every profession. AI may follow a similar path.

This does not mean everyone needs to understand how large AI models are trained. It means people increasingly need to understand what AI can do within their profession, where it can help, where it can fail and when human judgement remains necessary.

The durable skill is unlikely to be simply knowing how to give an AI instructions. Interfaces will become easier. The more durable capability is understanding a real problem deeply enough to use increasingly intelligent tools to solve it.

The cleaner understands sanitation. The farmer understands crops. The electrician understands electrical systems. The engineer understands infrastructure.

AI and robotics increase what those people can control.

Phase One: Save and Improve the Present

This book eventually travels far beyond Earth, but the Employment Bridge should begin here.

Before we build autonomous machines capable of maintaining infrastructure on the Moon or Mars, there are enormous problems to solve on Earth. Water systems need improvement. Polluted environments need restoration. Electricity networks need expansion and modernisation. Cities need better waste systems. Infrastructure needs inspection and repair. Natural disasters require better monitoring and response. Energy needs to become cleaner and more abundant. Healthcare, agriculture, transportation, manufacturing and construction all contain problems that AI and robotics could help address.

This is **Phase One** of the Employment Bridge: use AI and robotics to protect and improve the civilisation we already have.

This also serves another purpose.

Earth becomes our training ground.

Earth's Hard Problems Prepare Us for Space

A robot operating inside a sewer is obviously not the same as a robot operating on Mars. The environments are radically different. But both require machines that can perceive difficult surroundings, perform physical work, detect failures and operate where constant direct human intervention may be undesirable or impossible.

A robot inspecting remote water infrastructure teaches engineers about autonomous inspection. A disaster-response robot teaches navigation through damaged environments. An autonomous mining machine teaches excavation and material handling. A robotic repair system teaches fault detection and physical maintenance. An AI coordinating hundreds of machines teaches fleet management, permissions, scheduling and recovery.

Each capability has terrestrial value before it has extraterrestrial value.

The progression could therefore become:

Solve difficult problems on Earth → develop better robots → increase machine autonomy → build remote autonomous infrastructure → adapt those capabilities for space

This connects saving the present directly with building the future. We do not need to choose between repairing Earth and preparing for an interplanetary civilisation.

Done correctly, the first can help us learn how to build the second.

Phase Two: Build Industries Before We Need Them

The next stage requires thinking beyond the next financial quarter.

If humanity wants highly autonomous construction, mining, computing and energy systems several decades from now, the required knowledge cannot suddenly be created when we decide we need it.

We need robotics engineers, materials scientists, power engineers, software and platform engineers, network specialists, cybersecurity professionals, manufacturing experts, autonomous-systems researchers, construction specialists, environmental engineers and technicians capable of maintaining increasingly complex machinery.

The workforce itself is therefore part of the infrastructure.

A company laying off experienced technical employees because AI can perform some of their current tasks may be making a rational decision about today's cost structure. But at the level of the wider economy, allowing large amounts of technical capability to become idle while enormous engineering problems remain unsolved would be a failure of imagination.

Governments, universities, companies and research institutions therefore have a role in creating longer-term research and development programmes. Not every programme will succeed. Not every technology will become commercially viable.

But technological civilisation cannot prepare for fifty years from now if almost every investment decision is judged only by what improves next year's productivity.

Phase Three: Expand the Map

Over much longer timescales, the physical boundaries of economic activity could begin expanding beyond Earth.

A mature space economy would require energy, communications, robotics, transportation, mining, construction, manufacturing, maintenance and computing. Much of the physical work may initially be performed by robots because humans are fragile in space. But robotic industry does not eliminate the need for human knowledge.

Machines must be designed, manufactured, tested and secured. Their failures must be investigated. Their objectives must be established. Infrastructure must be planned. Supply chains must be managed. Laws and standards must be developed. New generations of technology must continually be invented.

Many people working in that economy may never leave Earth.

A future network engineer could manage communications spanning Earth, orbital infrastructure and the Moon. A software engineer could develop control systems for machines operating millions of kilometres away. A robotic technician could diagnose equipment remotely while local machines perform the physical repair.

Space employment does not mean everyone becomes an astronaut.

It means the environment in which human knowledge is applied becomes larger.

From a Finite Frontier to an Expanding Economy

This expansion could have consequences far beyond employment.

Many of today's economic and political pressures exist within a civilisation whose physical economy is concentrated almost entirely on one planet. Countries compete for energy, minerals, technology, investment, industrial capacity and strategic advantage. Within countries, people compete for employment, housing, education and economic security.

Expanding beyond Earth would not automatically eliminate these problems. Poverty is not caused only by a shortage of physical resources. Inequality can result from how wealth and opportunity are distributed. Financial crises can emerge from debt, markets, institutions and human behaviour. Racism and discrimination cannot be mined away from an asteroid. Political competition would not suddenly disappear because humanity reached the Moon or Mars.

But an expanding space economy could gradually change one condition beneath some of these pressures: **the physical frontier of economic activity would no longer end at Earth.**

This matters because AI is often discussed inside a fixed economic model. We imagine approximately the same industries producing approximately the same goods and services, except that machines perform more of the work.

In that model, increasing automation can appear mainly as a competition between people and machines.

An expanding economy creates another possibility.

AI and robotics could help civilisation build things that were previously too expensive, dangerous or difficult to attempt: environmental restoration on Earth, vastly larger energy systems, orbital infrastructure, lunar industry, autonomous mining, planetary communications and, much later, infrastructure around other worlds.

The question changes from:

How do we divide a fixed amount of work?

to:

How much more useful work can civilisation become capable of doing?

Competition Could Gradually Include More Cooperation

An interplanetary economy could also create new forms of international interdependence.

Building such an ecosystem would require an extraordinary combination of capabilities: launch systems, advanced semiconductors, telecommunications, AI, robotics, energy systems, mining, materials science, manufacturing, medicine, construction, logistics and a very large skilled workforce.

Different countries already possess different strengths in these areas. The supply chain behind a modern computer or spacecraft can cross many national borders before the final system exists.

A much larger space economy could deepen that interdependence.

One country might contribute advanced manufacturing, another launch capability, another materials processing, while others contribute software, engineering, scientific research, construction, operations and skilled human resources. Developing economies could participate not merely as consumers of technology but as contributors to increasingly complex international supply chains.

This does not mean national competition would disappear. Countries may continue to protect strategic technologies, compete for influence and attempt to maintain independent capabilities.

Competition and cooperation can exist at the same time.

The more interesting possibility is that the enormous scale and complexity of building infrastructure beyond Earth could make cooperation increasingly valuable even between countries that continue to compete in other areas.

The economic incentive gradually becomes not simply:

Who controls the largest share of a fixed system?

but also:

How much larger can the system become if different capabilities work together?

Extraterrestrial Resources Could Change the Scarcity Equation

Over still longer periods, the resource base itself could begin to expand.

The Moon, asteroids and other bodies contain enormous quantities of material. Accessing those resources economically would be extraordinarily difficult and would require mining systems, energy, processing equipment, transportation and infrastructure that largely do not exist today.

A resource is not economically useful merely because it exists somewhere in space.

But if those capabilities eventually develop, civilisation could begin obtaining some materials without relying exclusively on Earth's environment.

That could gradually reduce pressure from selected forms of terrestrial mining and resource-intensive industry. Some industrial activities that do not require Earth's biosphere could eventually be performed elsewhere, allowing Earth to be treated increasingly as the biological centre of civilisation—a place where land, water and ecosystems are valued not merely as inputs into industrial production, but for food production, biodiversity, forests, cities, recreation and environmental restoration.

This does not mean manufacturing suddenly disappears from Earth. Nor does it mean extraterrestrial materials automatically make every product cheaper. Advanced processors, for example, depend on extraordinarily sophisticated fabrication equipment, knowledge and supply chains, not simply access to raw materials.

The transition would be gradual.

But over generations, greater access to energy, materials and automated productive capacity could expand the amount of infrastructure civilisation is capable of building.

The opportunity can therefore be expressed more carefully as:

Greater resources + greater productive capacity + responsible institutions → the possibility of reducing some forms of scarcity

The final part matters.

Abundance does not guarantee fairness. A civilisation capable of harvesting enormous resources could still distribute the benefits badly. Technology creates possibilities; governance determines how many of those possibilities become broadly shared progress.

From a Vicious Loop to an Expansion Loop

There is a risk that AI produces a narrow economic loop:

Automate existing work → reduce workforce → lower costs → automate more existing work → reduce workforce again

At the level of an individual organisation, parts of this cycle may sometimes be unavoidable.

But civilisation does not have to make that its only loop.

We can build an expansion loop:

Automate tasks → increase productivity → release human capability → train people → attack unsolved problems → create new industries → develop better technology → attempt larger projects → create new work

The difference is ambition.

If the total amount of useful activity we attempt remains fixed, increasingly capable machines can make human labour appear progressively less necessary.

If the scale of our objectives expands with our capabilities, automation can become one of the mechanisms that allows civilisation to attempt things previously considered too expensive, dangerous or difficult.

And if the physical frontier of civilisation eventually expands as well, the economy itself need not remain permanently confined to the productive boundaries of one planet.

Billions of Brains Are an Asset

Global population growth is expected to slow substantially over the long term, although the exact peak and future trajectory remain uncertain. Even so, humanity is likely to remain a civilisation of billions of people.

It is easy to combine a population of that scale with increasingly capable AI and ask how we will find enough jobs for everyone.

There is another way to see the same future:

billions of people represent billions of human minds.

Not everyone needs to become a scientist, programmer or spacecraft engineer. A civilisation operating enormous technological systems requires many forms of intelligence and practical capability: engineering, construction, medicine, maintenance, logistics, education, administration, manufacturing, environmental management, governance, design and countless forms of local knowledge.

People also possess different abilities. Some will invent new systems. Some will operate them. Some will repair them. Some will organise people. Some will teach. Some will care for others. Some will perform practical work that remains difficult to automate.

There will always be differences in skill, motivation and opportunity. Technology cannot guarantee employment for every person or eliminate every social problem.

But a population should not be viewed simply as a labour cost waiting to be automated away.

Human capability is itself an infrastructure resource.

A civilisation trying to repair and improve infrastructure across Earth—and eventually operate increasingly sophisticated systems beyond it—may discover that its problem is not having too many humans. Its challenge may be developing enough capable people to manage everything it has become able to build.

The Employment Bridge May Take Generations

We should therefore think about the Employment Bridge across several timescales.

Now, people learn to work effectively with increasingly capable AI.

Next, AI and robotics expand into environmental restoration, sanitation, water, energy, infrastructure, manufacturing, healthcare, agriculture and other large physical problems.

Later, increasingly autonomous machines allow civilisation to undertake work that is too dangerous, remote or expensive for humans to perform directly.

Much later, those capabilities help support orbital infrastructure, lunar industry, planetary computing and eventually a larger Interplanetary Cloud.

There is no guarantee that every worker will move smoothly through this transition. Some occupations will shrink. Some will disappear. New jobs will not necessarily appear in the same location or at the same moment. Retraining will not solve every problem. Technological change will not be painless.

Likewise, expanding the resource frontier does not guarantee the disappearance of poverty, inequality, discrimination, financial instability or international conflict.

But there is an enormous difference between allowing disruption to determine the future accidentally and deliberately building pathways toward the opportunities technology makes possible.

AI Should Expand Human Ambition

The deepest purpose of the Employment Bridge is therefore not simply retraining.

It is expanding what civilisation attempts to accomplish.

AI gives us additional intelligence. Robotics gives that intelligence physical capability. Humans provide objectives, judgement, experience, creativity, responsibility and knowledge of the world in which those systems operate.

There are polluted waterways to restore, sanitation systems to improve, electricity networks to modernise, diseases to treat, infrastructure to rebuild, environmental risks to manage, scientific questions to investigate and technologies still to invent.

Farther ahead lies an entire Solar System in which humanity has barely begun to operate.

The important question is therefore not simply whether humans or AI will have the jobs.

A more useful question is:

What becomes possible when humans, AI and robots work together—and are we ambitious enough to attempt it?

Today, we fear AI partly because we can see the jobs it may disrupt more clearly than the industries it could help us build.

The Employment Bridge is intended to close that gap.

It begins with protecting and developing the worker facing disruption today. It continues by using AI and robotics to solve problems on Earth that have remained neglected for too long. It extends through new industries, new skills and new infrastructure. Over generations, it may reach beyond Earth itself.

If that expansion eventually gives civilisation access to greater energy, materials and productive capacity, some of the physical scarcity pressures that shape today's economy could also begin to change.

But technology alone cannot decide what kind of civilisation emerges from that abundance. That remains a human responsibility.

The far end of the bridge is therefore not one particular profession, one industry or even one planet. It is a civilisation that refuses to respond to greater intelligence by making human ambition smaller.

Instead, it uses greater intelligence to make its ambitions larger.

AI should not merely help us do today's work with fewer people. It should help billions of people attempt what humanity could never accomplish before.

PART IV THE DATA EXPLOSION

AI is not only a compute problem; it is a storage, movement and resilience problem

Closing Part III

Part III connected environmental responsibility, automation and employment into one transition. AI and robotics should not advance by making either Earth or human capability disposable. Their greater value may lie in allowing civilisation to attempt projects that were previously too difficult, dangerous or expensive. But expanding intelligence creates another challenge: expanding information. AI does not merely process the world's existing data. It continuously creates more. Part IV turns to the storage, movement and resilience consequences of that explosion.

What PART IV Covers

AI is usually described as a consumer of enormous datasets, but increasingly it is also a producer of data. AI systems, robots, sensors, scientific instruments and digital infrastructure can generate vast quantities of logs, models, images, decisions, telemetry and synthetic information. This Part examines what happens when computation expands together with the information it creates. Not every byte deserves permanent storage, but valuable data requires protection, replication and sometimes geographic separation. The discussion therefore moves from conventional backup toward planetary resilience: a future in which selected scientific, cultural and technological information could eventually be replicated beyond Earth rather than keeping every important copy on a single planet.

Chapter 13

AI Does Not Just Consume Data. It Creates It.

Artificial intelligence is usually described as hungry for data. Models train on enormous datasets, and AI systems depend on text, images, code, sensor records and other digital information. But that is only half of the story. AI does not merely consume data. It creates it. It generates text, images, code, audio, video, summaries, logs, decisions, synthetic datasets and operational records. As AI becomes embedded in businesses, science, transport, robotics and infrastructure, machines themselves become large-scale producers of information. Creating information is not free. Data has to be stored somewhere. Important information is usually copied. Some needs backup. Some must be retained for legal, scientific or operational reasons. Some has to move across networks. And some should probably be deleted. The future of AI is therefore not only about how much computation civilisation can perform. It is also about how much information we can sensibly create, move, preserve and protect.

Civilisation Is Becoming a Recording System

Humanity was producing enormous quantities of data long before generative AI. Payments create transaction records. Websites create logs. Hospitals create medical records. Factories record machine states. Scientific instruments produce datasets. Cars contain cameras, radar and other sensors. Satellites continuously observe Earth and space. Now add AI and increasingly autonomous machines. An AI customer-service system may create conversations, summaries and decision records. An infrastructure agent may retrieve logs, call APIs, diagnose failures and record the actions it performed. A robot may continuously record its environment, movement, mechanical condition and decisions. What appears to a human as one AI action may therefore create many layers of information behind the scenes. Some of those records are important because they provide accountability. If an AI controlling an important system makes a serious mistake, investigators may need to understand what happened and why. AI therefore adds another layer to a civilisation that was already becoming a giant recording system.

AI Can Create Data to Train AI

AI also produces **synthetic data**: information generated artificially rather than directly collected from real events. Consider an autonomous vehicle. Collecting real examples of every dangerous driving situation would be difficult and sometimes unsafe. Simulations can instead create rare scenarios such as poor visibility, unexpected pedestrians or unusual traffic behaviour. The system can then learn from those artificial situations. This creates an unusual loop:

Data trains AI → AI creates synthetic data → synthetic data helps train future AI.

Because machines can generate such information at enormous scale, data production no longer depends mainly on human activity.

AI Creates Hidden Data Too

The visible output of an AI system is only part of its data footprint. Models themselves contain large collections of numerical parameters learned during training. Training systems may preserve checkpoints, allowing work to resume from an earlier stage rather than restarting an expensive training process. AI search systems can also create embeddings—numerical representations that help systems find information by meaning or similarity. The original documents remain, while embeddings and search indexes create additional layers around them. One dataset can therefore produce:

Original information → AI representations → indexes → operational logs → backups and replicas.

The user may see only one answer. The infrastructure may be maintaining several layers of data behind it.

Data Is Rarely Stored Only Once

Suppose an organisation has one petabyte of important information. That does not necessarily mean it needs only one petabyte of physical storage. Production systems may maintain replicas for reliability. Backups preserve earlier states. Archives retain historical information. Analytics systems create transformed copies. AI systems may create embeddings and indexes. One petabyte of useful information can therefore require several petabytes of physical storage. This duplication is not automatically waste. Storage devices fail. Software can corrupt information. People delete files accidentally. Cyberattacks can destroy or encrypt data. Entire facilities can lose power or suffer physical damage. Reliable systems therefore create copies because valuable information needs protection. But as civilisation creates more data, every decision to preserve another copy also creates additional infrastructure.

Data Has Gravity

As datasets become larger, moving them becomes increasingly difficult. Engineers sometimes describe this as data gravity: large datasets tend to attract computing toward them because moving the computation can be easier than moving the entire dataset. Imagine a library containing ten million books. If a researcher needs to study those books, sending the researcher to the library may be easier than transporting the whole library to the researcher. Large digital datasets can create a similar effect. This becomes particularly important beyond Earth. A lunar facility will generate lunar data. Mars will generate Martian data. Space telescopes will generate astronomical data. Robotic fleets may continuously produce sensor and operational information. Sending every raw byte back to Earth may not make sense. Sometimes: the computation should move toward the data rather than the data moving toward the computation.

The Future Robot Is a Data Factory

Consider a future lunar mining robot. It may contain cameras, navigation sensors, drilling instruments, power monitors and systems measuring its own mechanical health. It needs much of this information to operate safely and autonomously. Multiply that by thousands of robots. A robotic industrial system becomes not only a consumer of AI but a continuous data factory. Some information may need immediate processing. Some may be useful for maintenance. Some may have scientific value. Some may be retained for improving future machines. And much of it may eventually become worthless. That means storage architecture has to be designed alongside robotic architecture. Where should the information be processed? What stays on the machine? What moves to a nearby data centre? What reaches Earth? What deserves long-term preservation? Those decisions become part of the architecture itself.

We Need Intelligent Forgetting

Digital systems have developed a habit that biological memory does not share very well: keeping almost everything. When humans generate information slowly, that may be manageable. When machines can generate information continuously with almost no human effort, storing everything indefinitely becomes increasingly irrational. We therefore need intelligent forgetting. Information can be evaluated according to its purpose. Is it required for current operations? Does law require it to be retained? Is it scientifically valuable? Is it needed for security or accountability? Could it help explain an important AI decision? Can it be compressed? Has the important knowledge already been extracted from the raw data?

If none of those reasons applies, deletion may be the correct engineering decision.

AI itself could help make these classifications under policies established by humans. The system creating the data could also help manage its lifecycle.

Not All Data Deserves the Same Storage

Engineers commonly distinguish between **hot, warm and cold data**. Hot data is needed frequently and quickly. Warm data is accessed less often. Cold data may be rarely accessed but still worth preserving. A future planetary infrastructure could apply the same principle. Current operational information on Mars might remain in fast local storage. Older logs could move into slower archives. Important scientific discoveries might be transmitted to Earth. Civilisation-critical information could receive additional copies elsewhere. Not everything needs to occupy the fastest storage. And not everything needs to exist on every world.

Distance Makes Selection More Important

Planetary distance makes another familiar technique increasingly valuable: **compression**. Imagine a Mars robot generates hundreds of gigabytes of sensor data. Scientists on Earth may not need every raw camera frame immediately. Local AI could analyse the information first and prioritise unusual events, scientific discoveries, compressed images and important summaries. The complete raw dataset could remain locally available when necessary. This is an extension of edge computing: processing information close to where it is created. On Earth, edge computing can reduce latency and network traffic. Across planets, it becomes much more important because bandwidth is limited and distance cannot be removed.

AI Could Become the Librarian of the Solar System

As machine-generated data becomes too large to move indiscriminately, AI may increasingly help decide what deserves priority. A future lunar science facility might classify information according to rules such as:

Urgent operational information → transmit immediately.

Routine telemetry → summarise and aggregate.

Large raw datasets → process and archive locally.

Duplicate or temporary information → delete when policy permits.

Civilisation-critical knowledge → preserve with additional copies.

Humans, organisations and law would establish the retention policies. AI would help perform the enormous operational task of applying them. This creates an interesting symmetry. AI contributes to the data explosion, then AI helps civilisation manage the data explosion. The technology increases complexity and simultaneously becomes one of the tools required to control that complexity.

Storage Is Also a Physical Problem

Digital storage may appear almost weightless to the person saving a file, but the information ultimately exists in physical infrastructure. Storage devices consume electricity. Controllers and networks consume electricity. Hardware must be manufactured. Data centres require cooling. Replication and backup require additional equipment. The environmental question surrounding AI is therefore not only: How much energy does AI consume while computing? It is also: How much infrastructure do we maintain to preserve everything AI and civilisation create? This is another reason why intelligent deletion, compression and tiered storage matter.

What Is Worth Preserving?

Eventually, the engineering question becomes a philosophical one. Humanity is producing more information than any previous civilisation. But data is not the same as knowledge. A million raw sensor readings are data. A reliable scientific conclusion extracted from them may become knowledge. Both may sometimes deserve preservation, but they do not automatically deserve identical treatment. What should humanity preserve for a hundred years? Scientific discoveries? Medical knowledge? Engineering designs? Cultural works? Historical records? Genetic information? Software? AI models? Different organisations and individuals will answer differently. But a civilisation-level archive would require prioritisation. A future extraterrestrial archive intended to protect important human knowledge would probably not contain every piece of digital content civilisation ever produced. Its purpose would be to preserve what future civilisation may genuinely need.

From Data Storage to Civilisation Memory

Most organisations think about backup at organisational scale. Protect customer records. Restore business systems. Preserve financial information. A civilisation operating across multiple worlds could eventually ask a larger question: What information would humanity need to preserve if a major disaster destroyed part of its technological infrastructure? Scientific knowledge, medicine, agriculture, manufacturing, engineering, software, cultural history and other foundational information could become part of what we might call civilisation memory. Today, almost all meaningful copies of humanity's digital knowledge share one physical fact: they are on Earth. That is entirely reasonable because Earth is currently where our civilisation and almost all of our computing infrastructure exist. But if humanity eventually establishes permanent computing and storage infrastructure beyond Earth, another possibility appears. Important information could be classified according to scale:

Local operational data → kept where it is needed.

Planetary important data → protected within that world.

Cross-planetary critical data → given another planetary copy.

Civilisation memory → selected for very long-term preservation.

The objective would not be to copy every byte everywhere. It would be to decide which information is important enough to survive beyond a single location—and eventually beyond a single planet.

The Future Is Not Infinite Storage

It would be easy to conclude that AI simply requires humanity to build larger and larger data centres forever. I think the better conclusion is almost the opposite. An advanced technological civilisation should become more selective about information, not merely better at hoarding it. The goal is not to preserve every byte civilisation ever creates. It is to preserve the information required for operations, accountability, science, culture and future understanding while avoiding unlimited digital waste. Once we make that distinction, a new architectural question appears.

If humanity eventually identifies a relatively small subset of information important enough to protect for generations—or even against planetary-scale disasters—

where should those copies live? Today we protect important data using another server, another data centre or another geographic region. A future civilisation may eventually add another level: another world. That takes us from the data explosion to the next architectural problem: From Backup to Planetary Resilience.

Chapter 14

From Backup to Planetary Resilience

In the previous chapter, we saw that AI creates a second problem beyond computation: **data must survive**. Modern civilisation already protects important information by keeping copies across servers, data centres and geographic regions. As our dependence on digital systems increases, however, we can eventually ask a larger question: Should every critical copy of humanity's digital knowledge remain on Earth? This is not an argument for sending today's corporate backups to Mars. For the foreseeable future, terrestrial regions provide vastly cheaper, faster and more practical disaster recovery. The argument concerns a civilisation operating decades or centuries from now, when permanent infrastructure may exist in orbit, on the Moon and perhaps eventually on Mars. At that point, planetary separation could become another layer of resilience.

Resilience Begins with Failure Domains

A **failure domain** is a group of systems that can be affected by the same failure. Two copies of a database on different disks may survive one disk failing, but not destruction of the server. Copies on different servers provide greater protection, but perhaps not against a building fire. Another copy in a distant data centre protects against a wider range of events. Cloud architecture already extends this principle through machines, availability zones and geographic regions. The progression is conceptually simple:

Server → Data Centre → Availability Zone → Region → Continent → Planet

We do not need the final layer today. But a civilisation with permanent off-world infrastructure might eventually decide that its most important information deserves it.

Earth Would Remain the Primary Location

Earth would continue to host most human data for a very long time. Almost all users, businesses and institutions are here. Keeping information near them provides low latency, enormous network capacity and easier maintenance. Sending ordinary transactions to another planet would make little sense. A planetary backup would serve a completely different purpose. Imagine a carefully selected archive containing scientific knowledge, medicine, engineering designs, agricultural information, important software, cultural records, genomic resources and perhaps selected AI models. Most copies would remain distributed across Earth. An additional off-world copy would exist for a different reason: civilisation-level resilience.

Why the Moon Is the Natural First Step

The Moon is, on average, approximately **384,400 kilometres from Earth**. A signal travelling at the speed of light takes about **1.28 seconds one way**, before additional processing and routing delays. That is inconvenient for highly interactive computing but entirely manageable for many archival applications. More importantly, the Moon has a different physical risk profile. An Earth-based wildfire cannot reach it. A terrestrial flood cannot reach it. A collapse of an Earth electrical grid cannot directly shut down lunar infrastructure. This does not mean the Moon is inherently safer. Lunar infrastructure introduces completely different problems: radiation, extreme temperature variation, abrasive dust, power generation, communication and difficult maintenance. The advantage is not absence of risk. It is diversity of risk. In resilience engineering, that can be extremely valuable.

A Planetary Backup Does Not Need Every Byte

Humanity would not need to copy the entire Internet to the Moon. Chapter 13 introduced intelligent forgetting: the idea that advanced digital systems should distinguish between information worth preserving and information that can eventually be discarded. A civilisation archive could apply the same principle. Think of it as something loosely comparable to a seed vault—but for knowledge. Scientific literature, medicine, mathematics, engineering, agriculture, maps, software, manufacturing knowledge and important historical and cultural records might deserve long-term protection. AI could help organise and update such an archive, but deciding what deserves inclusion would ultimately remain a human and institutional responsibility.

Recovery Does Not Require a Perfectly Current Copy

Backup engineering already provides a useful concept called the **Recovery Point Objective**, or RPO. RPO describes how much recent information can acceptably be lost following a failure. A financial system may require extremely recent copies because losing transactions could be serious. An archive of scientific literature may tolerate a much longer delay because its contents change relatively slowly. A future resilience hierarchy could therefore look conceptually like:

Primary Earth systems → Earth replicas → regional backups → lunar archive → eventual Martian archive

The farther away the copy, the less frequently some categories of information may need to be updated. Planetary resilience therefore does not require making interplanetary communication behave like a terrestrial network. We design the system around the delay.

Mars Changes the Meaning of Backup

Mars makes this principle much more obvious. Earth and Mars continuously move around the Sun, so the distance between them changes enormously. Depending on their positions, one-way communication takes minutes rather than milliseconds. Imagine clicking Save on Earth and requiring a database on Mars to confirm the transaction before the application can continue. That architecture would be unusable. A Martian archive would therefore have to operate asynchronously. Earth continues operating. Selected changes accumulate and are transmitted when possible. Mars gradually incorporates them into its own archive. The Martian copy is therefore somewhat behind Earth. For civilisation-level disaster recovery, that may be completely acceptable.

The Speed of Light Cannot Be Engineered Away

This places an important physical boundary around the Interplanetary Cloud. Radio, microwaves and laser light are all electromagnetic radiation and travel through vacuum at approximately the speed of light. Laser communication may provide much greater data capacity than some radio systems, but it cannot make information travel faster than light. Adding relay stations between Earth and Mars therefore cannot turn minutes of propagation delay into milliseconds. The relays solve a different problem. They can improve continuity, storage, routing and resilience. This distinction is fundamental to the architecture proposed in this book.

Communication Becomes Store-and-Forward

Terrestrial networking usually works best when an end-to-end path is continuously available. Space cannot always provide that. A spacecraft may move behind a planet. Antennas can lose alignment. Equipment can fail. Communication windows may be intermittent. Solar geometry can disrupt some links. Space networking therefore uses concepts such as Delay/Disruption-Tolerant Networking (DTN). One important standard within this approach is the Bundle Protocol Version 7, defined by the Internet Engineering Task Force in RFC 9171 for networks that may experience intermittent connectivity and large or variable delays. Instead of requiring an uninterrupted path from sender to destination, information can be stored temporarily at an intermediate node and forwarded when another useful communication opportunity becomes available.

Imagine:

Earth → Earth Orbit → Deep-Space Node → Mars Orbit → Mars

A block of information reaches an orbital node and is stored safely. When the next connection becomes available, it continues onward. The information gradually progresses through the network.

From Relay to Mini Data Centre

This is where I extend the concept into the architecture proposed in this book. Instead of treating important intermediate nodes only as communications relays, imagine gradually giving some of them: communication + storage + computing + autonomous management A node could receive data, verify it, retain it temporarily and forward it when the next link becomes available. It could also cache useful information, perform local processing, monitor its own hardware and help determine the best available route. Data would therefore no longer exist only at the source and destination. For periods of time, it would also live inside the network. The network begins to behave partly like a communications system and partly like a distributed storage and computing platform. This is one of the core ideas behind the Interplanetary Cloud.

The Network Is Moving

There is an important complication. We cannot place a straight row of fixed towers between Earth and Mars. Earth moves around the Sun. Mars moves around the Sun at a different speed. The Moon orbits Earth. Every relay placed in space must itself follow an orbit or another controlled trajectory. The geometry continually changes. The architecture is therefore not:

Earth → Fixed Relay 1 → Fixed Relay 2 → Fixed Relay 3 → Mars

Instead, different nodes become useful at different times. Some may orbit Earth or the Moon. Some may orbit Mars. Others could follow independent heliocentric orbits around the Sun. Certain orbital regions associated with Lagrange points may also provide useful locations for particular infrastructure. The important principle is that the network must understand where its nodes will be, not merely where they are now. Its routes therefore change with the geometry of the Solar System. Rather than a fixed chain, the Interplanetary Cloud becomes a moving, predictive network of communication, storage and computing nodes. We will examine that networking architecture in greater detail later.

A Relay Is More Than a Signal Booster

Calling these nodes simple **signal boosters** would also be misleading. A booster or repeater primarily regenerates or amplifies a signal. A more capable interplanetary node could receive information, verify it, store it, determine where it should travel next and later generate a new transmission toward another node. It therefore behaves more like a combination of: relay + router + storage system + computer Future routing decisions could consider predicted orbital geometry, available bandwidth, energy, congestion, reliability and urgency. AI could help calculate those routes because the topology of the network continually changes.

From Relay to Infrastructure Hub

Once a node contains substantial power, communications, storage and computing, another possibility appears. Why use it only for networking? Some larger nodes might eventually provide navigation services, spacecraft monitoring or support for nearby robotic operations. Much later, selected locations might develop into more substantial infrastructure hubs. Not every node would do everything. Some could specialise in communications. Others in computing or storage. Some might combine several functions. The Solar System would not contain a fixed highway lined with stationary service stations. A better analogy is a system of moving ports whose future positions can be predicted. Transport and communication routes would be planned around that changing geometry.

Distance Creates the Need for Local Intelligence

Once Earth, the Moon, orbital infrastructure and Mars contain independent computing systems, planetary resilience becomes more than backup. A lunar research facility could process its own scientific data. Mars could operate specialised AI using locally generated information. Orbital observatories could analyse images before transmitting selected results to Earth. Each location develops a degree of digital autonomy. That autonomy is necessary because distant systems cannot depend on Earth for every decision. The relationship becomes: Distance creates latency. Latency requires autonomy. Autonomy requires intelligence. Intelligence requires local computing. The farther infrastructure moves from Earth, the stronger the case becomes for local computing and storage. The Interplanetary Cloud is therefore not one enormous terrestrial-style cloud stretched across the Solar System. It is better understood as a federation of local computing environments connected by delayed, changing communication paths.

Earth Is the Beginning, Not the Only Node

For most of human history, preserving knowledge meant protecting books, manuscripts and physical records.

The digital age changed preservation into replication across machines. Cloud computing extended that principle across geographic regions. A mature space civilisation could eventually extend it across worlds. This does not mean abandoning Earth. Earth would remain the centre of human civilisation for the foreseeable future. Planetary resilience means something different: as our civilisation becomes capable of operating beyond Earth, some of its most important knowledge could eventually become more physically distributed than ever before. The progression is straightforward: Back up the machine. Then the building. Then the region. Eventually, perhaps, the planet. But the moment we seriously consider that final step, conventional cloud engineering encounters a completely new problem. Our data centres are no longer separated by kilometres. They may eventually be separated by hundreds of millions of kilometres. The next challenge is therefore not simply where we store information. It is: How do we communicate with it across the Solar System?

Before putting data centres in orbit or on the Moon, why not distribute AI and storage across billions of phones, laptops, vehicles and other devices already on Earth?

Before asking why civilisation might someday build computing infrastructure beyond Earth, we should ask a simpler question. *Why build more data centres at all? Billions of people already carry powerful computers in their pockets. Phones, laptops, vehicles and other devices contain processors, storage and increasingly dedicated AI hardware. Why not distribute civilisation's AI across these machines instead of concentrating so much computation inside enormous facilities?*

To a significant extent, we should. This is already beginning to happen. Smaller language models can operate directly on consumer devices, allowing some AI requests to be processed without sending everything to a remote server. Local processing can reduce communication delay, improve privacy, continue working during some network outages and reduce unnecessary traffic to central infrastructure. As hardware improves, much more routine AI could move toward the device where it is actually needed.

But a handheld device is not a miniature data centre. It has limited memory, storage, battery capacity and ability to remove heat. The largest AI systems may require far more computing and memory than a consumer device can economically provide. Training future generations of AI creates an even larger problem because enormous numbers of specialised processors may need to exchange information rapidly while consuming large quantities of electricity. A phone may be excellent at running a smaller already-trained model, but that does not mean billions of phones can conveniently replace a tightly connected training cluster.

Storage creates similar trade-offs. Consumer devices could potentially hold encrypted fragments, caches or additional copies of selected information. But they are unreliable as permanent infrastructure. Phones are switched off, disconnected, damaged, lost and replaced. People own them, organisations have different security requirements, countries apply different laws, and important information would need enough redundancy to survive devices continually leaving the network. Turning billions of personal devices into a global storage system would therefore create a massive distributed-systems problem of its own.

There is also little reason for every device to contain every piece of information. A personal AI may need knowledge and models relevant to its owner, while a scientific institution may need petabytes of specialised observations and a global AI service may require much larger models. Replicating all of those resources onto billions of devices would itself consume enormous quantities of storage and hardware.

The more realistic future is therefore hierarchical rather than centralised or completely distributed:

Device → Local Edge → Terrestrial Compute Region → Orbital Region → Planetary Region

The device performs work that benefits from privacy, immediacy and local context. Edge infrastructure supports nearby machines, factories, vehicles or communities. Terrestrial data centres provide large, shared models, databases, training systems, backups and high-performance computing. Much later, orbital and planetary regions could perform specialised workloads, provide additional storage and resilience, support extraterrestrial activity and absorb some forms of infrastructure growth that no longer need to remain on Earth's surface.

Efficiency Delays Expansion; It Does Not Necessarily End It

This distinction matters because technological efficiency does not automatically reduce civilisation's total demand for computing. A future processor may perform today's AI calculation using a fraction of today's electricity. A phone may run locally a model that once required a server. Storage density may increase enormously.

But increased efficiency can also make AI cheaper and therefore allow it to be used in far more places.

One AI assistant becomes billions of personal assistants. A few industrial robots become millions of autonomous machines. Scientific AI expands into medicine, materials, climate, engineering and planetary exploration. Vehicles become intelligent computing platforms. Cities operate enormous sensor networks. Robots generate continuous telemetry. New AI systems are trained, tested, specialised, copied and preserved. Increasing intelligence creates new applications, and those applications create additional computation and data.

The long-term equation may therefore become:

More efficient computing × vastly more uses = continuing infrastructure growth

This is why the argument for the Interplanetary Cloud does not depend on today's processors remaining inefficient.

If humanity can satisfy future computational demand through better terrestrial technology with minimal environmental impact, that should be welcomed. There is no reason to launch infrastructure into space merely because space exists. Earth will remain the most practical computing region for enormous amounts of human activity.

But this book is considering a civilisation evolving across decades and centuries. If AI continues becoming more capable, robots multiply, scientific datasets expand, autonomous infrastructure spreads and eventually economic activity extend beyond Earth, the total amount of computation and information civilisation maintains could become vastly greater than it is today.

At that scale, the question changes from:

Can we fit another data centre somewhere on Earth?

to:

Why should every layer of an expanding technological civilisation remain physically concentrated on one planet?

Earth Should Not Carry Every Layer Forever

Land on Earth has value beyond its price as industrial property. It supports cities, agriculture, forests, ecosystems, recreation and human communities. Electricity, water, minerals and manufacturing capacity also have competing uses. The objective should therefore not be to move data centres into space because terrestrial land has suddenly become unavailable. There is still substantial room to improve terrestrial computing.

The longer-term principle is different.

Civilisation should use Earth's resources intelligently while gradually learning how not to make Earth the permanent physical host for every future expansion of computation, storage, energy generation and industry.

Orbital computing might eventually serve spacecraft and Earth-observation systems where the data already originates in space. Lunar computing could support lunar science, industry and settlements rather than transmitting every operation back to Earth. Extraterrestrial archives could provide additional physical separation for selected knowledge. Future AI training or scientific workloads that do not require immediate interaction with people might eventually operate where suitable energy, cooling, communications and materials are available.

In this architecture, space does not replace terrestrial computing.

It adds another layer.

Phones do not replace data centres. Edge systems do not replace cloud regions. Orbital infrastructure does not replace Earth. Lunar infrastructure does not replace orbital infrastructure.

Each layer reduces the amount of work unnecessarily pushed onto another.

The Interplanetary Cloud is therefore not an argument for abandoning the terrestrial cloud. It is the long-term extension of the same architectural idea that created edge computing in the first place:

Put computation and data where they can be handled most appropriately.

Today that decision may be between a phone and a data centre.

One day, it may also be between worlds.

PART V THE FIRST STEPS OFF EARTH

From terrestrial cloud regions to orbital and lunar infrastructure

Closing Part IV

Part IV expanded the problem from computation to information itself. A highly digital civilisation must decide not only how much data to create, but what to retain, where to store it and how widely to replicate what truly matters. Terrestrial infrastructure should remain the primary layer, but the logic of distributed systems suggests that increasingly distant copies could eventually provide additional resilience. That raises the next question: if computing and storage are ever to extend beyond Earth, where should the first nodes actually go? Part V begins that transition.

What PART V Covers

This Part takes the architectural principles of modern cloud computing and cautiously extends their geographic scale. Today's cloud distributes workloads among regions, availability zones, data centres and edge locations. A much later infrastructure could add planetary and orbital regions to that hierarchy. Earth would remain the principal computing centre, while orbital systems, the Moon and eventually Mars could perform specialised roles according to distance, energy, communication delay and local demand. The chapters examine why the Solar System can be thought of as a future compute geography and why orbital computing is the most plausible first experimental step. The Interplanetary Cloud is therefore presented not as a replacement for terrestrial computing, but as its long-term architectural extension.

Chapter 15

The Solar System as the Next Compute Region

Cloud computing changed how engineers think about geography. Instead of assuming that an application must run in one building, modern platforms distribute systems across locations. A service may operate in one city, maintain another copy elsewhere and keep disaster recovery in another region. A **cloud region** is a geographic area containing computing infrastructure, often divided into physically separated **availability zones** so that failure in one location does not necessarily disable the entire service.

Today, all major cloud regions are on Earth. The long-range question explored in this book is whether that must always remain true.

From Geographic Region to Planetary Region

A cloud engineer today might think of infrastructure as:

Global platform → Region → Availability Zone → Data Centre → Cluster → Server

A **cluster** is a group of computers working together so applications can share processing capacity and continue operating when individual machines fail.

Now cautiously extend the hierarchy:

Solar System → Planetary Region → Orbital Zone → Data Centre → Cluster → Compute Node

Earth becomes one planetary region. The Moon could eventually become another specialised region. Orbital facilities could provide edge computing and communications. Much later, Mars could become a substantially more independent region with its own computing, storage and AI systems.

This does not mean a future cloud provider will simply add *Mars* beside Sydney or Frankfurt in a region-selection menu. Planetary distances fundamentally change networking, operations and reliability. The cloud-region analogy simply provides a familiar engineering framework from which to begin.

The Solar System Compute Fabric

I call the mature long-range version of this architecture the **Solar System Compute Fabric (SSCF)**.

A compute fabric is a connected collection of computing, storage and networking resources operating as parts of a larger system. The SSCF is not one giant computer controlling the Solar System, one central AI, or a Dyson sphere surrounding the Sun. It is a conceptual network of independently operating compute regions distributed across Earth, orbit, the Moon, Mars and eventually other useful locations.

The important word is **independently**.

A lunar facility must continue functioning when communication with Earth is unavailable. A Mars facility must make many decisions locally because communication delay prevents real-time Earth control. An orbital node may need to store information until another communication path becomes available.

The farther civilisation expands, the more distributed its computing architecture must become.

Computing Should Move Only When the Workload Benefits

Earth will remain the easiest place to build most computing infrastructure for a very long time. It already has factories, fibre networks, skilled workers, electricity grids, transportation systems, industrial supply chains and mature data-centre infrastructure.

The argument for off-world computing is therefore **not** that Earth is becoming unusable or that data centres should simply be moved into space.

Off-world computing becomes useful when the workload itself benefits from being there: processing data generated in space, supporting lunar or Martian industry, operating autonomous robots, conducting scientific research, maintaining planetary archives or performing other activities that would otherwise depend excessively on communication with Earth.

The principle should therefore be:

Do not move computing into space merely because space exists. Move it when the workload benefits from being there.

This also means that many applications should remain on Earth. A person asking an AI assistant a question should normally receive an answer from relatively nearby infrastructure. Sending an ordinary interactive request to Mars would introduce minutes of unnecessary delay.

The SSCF is therefore **workload-aware**. Computation belongs where its physical, economic and operational requirements make sense.

The First Off-World Compute Is Already Here

In a limited sense, computing has already left Earth. Satellites, spacecraft, the International Space Station and Mars rovers all contain computers. These are not hyperscale data centres, but they demonstrate that digital systems can operate beyond Earth's surface for years.

A spacecraft computer performs navigation, communication, scientific processing and equipment control. Mars rovers already perform some local processing because waiting for Earth before every action would be impractical. Satellites can increasingly analyse information before transmitting it to the ground.

This is an early form of **orbital edge computing**.

Edge computing means processing information close to where it is generated rather than sending every raw piece of data to a distant central data centre. Consider an Earth-observation satellite capturing thousands of images. Many may be obscured by clouds. Instead of consuming scarce communication bandwidth transmitting everything, onboard computing could identify unusable images, detect wildfires or storms and prioritise important observations.

The satellite stops being merely a sensor.

It becomes a small computing node.

The same principle becomes much more important as activity expands toward the Moon and Mars.

Why would anybody pay to build this before there is a Moon/Mars economy?

The Commercial Bridge Starts Much Closer to Earth

The economic case for off-world computing does not need to begin with enormous data centres on the Moon or Mars. It can begin much closer to Earth, wherever processing information in space provides a practical advantage over sending everything to the ground.

Earth-observation satellites provide one example. Instead of transmitting every image first, onboard or nearby orbital computing can identify useful observations, combine information from multiple satellites and prioritise urgent data. Communication infrastructure creates another opportunity. As optical links increase the amount of information spacecraft can exchange, some orbital nodes could combine communication, storage and computing—routing traffic, temporarily caching data and processing information before forwarding it.

Spacecraft telemetry provides a third use. Satellites continuously generate information about their own condition and operation. Rather than transmitting every raw measurement immediately, local systems could identify anomalies, summarise routine information and give priority to data requiring attention. As satellite constellations become larger, the value of managing information closer to where it is generated could increase.

These systems would not yet constitute the Interplanetary Cloud. They would be its economic and technological seeds: useful infrastructure built first to solve problems near Earth, then gradually extended as activity moves toward the Moon and farther into the Solar System.

The Moon as the First Major Compute Extension

The Moon is the natural candidate for the first substantial off-world compute region because it is relatively close to Earth. Its average distance is about

384,400 kilometres, and light takes roughly 1.28 seconds to travel that distance one way.

That delay is noticeable but manageable for many activities.

A mature lunar computing environment could support scientific instruments, navigation, mining equipment, habitats, construction robots, communications and industrial systems. More importantly, once robots and people operate there continuously, the Moon itself becomes a major source and consumer of data.

Sending every raw sensor record, camera stream and operational log back to Earth would become increasingly inefficient. Local computing develops because local activity creates a reason for it.

The progression is therefore:

Activity → Data → Local computing → Larger local infrastructure

rather than:

Build enormous data centre → Hope someone eventually needs it

This distinction makes the concept economically more credible. Computing should follow economic and scientific activity rather than being constructed merely for technological spectacle.

A Lunar Region Would Be Different from an Earth Region

An Earth cloud region may consist of several large buildings connected by high-capacity fibre networks. A lunar region would operate under very different constraints.

Hardware transportation would initially be expensive. Radiation protection would be important, and abrasive lunar dust could damage equipment. Cooling would also remain a major constraint. As discussed for orbital computing, vacuum does not remove heat automatically, so lunar facilities would require carefully designed thermal systems and radiators. Maintenance would need to become highly robotic, while power availability would vary according to location and solar conditions.

The architecture would therefore probably grow incrementally: a scientific base, communications infrastructure, robotic operations, local storage and a small compute cluster. As the surrounding lunar economy expands, computing expands with it.

A mature lunar region would emerge from infrastructure demand rather than arrive fully formed.

Orbital Zones and Relays

Not every useful computing location needs to sit on a planetary surface. **Orbital zones** could provide intermediate layers of communication, storage and processing.

An Earth-orbit node might process satellite traffic. A lunar-orbit node could connect Earth with infrastructure on the Moon's far side. A Mars-orbit node could link surface facilities, arriving spacecraft and Earth. More capable nodes might temporarily store information or process it before forwarding it.

This creates a hierarchy:

Planetary surface systems → Orbital edge systems → Interplanetary network

The far side of the Moon demonstrates why this matters. Because the Moon is tidally locked to Earth, approximately the same lunar hemisphere always faces us. A facility on the far side does not have direct line-of-sight communication with Earth and therefore requires relay infrastructure positioned where both sides can communicate.

The relay does not make light travel faster. It solves a **geometry and connectivity problem**.

Future orbital nodes could extend this idea by combining communications with storage and computing.

Distance Requires Local Control

One of the most important differences between terrestrial and interplanetary computing is the loss of practical real-time central control.

Cloud platforms use a **control plane** to decide where workloads run, how resources are configured and which policies apply. On Earth, geographically separated systems can usually communicate quickly enough for extensive coordination.

Mars cannot operate that way.

A mature Martian compute region would require its own local control plane. If Earth communication becomes unavailable—or simply takes too long—Martian systems must continue functioning. Local networks route traffic. Local storage serves applications. AI allocates computing resources. Robots continue working. Earth receives updates later.

The architecture therefore resembles **cooperating local clouds** more than one terrestrial cloud stretched across space.

Distance creates natural computing tiers. Earth-to-Earth communication may occur in milliseconds or fractions of a second. Earth-to-Moon communication takes seconds. Earth-to-Mars communication takes minutes, with the delay varying as the planets move. More distant infrastructure experiences still greater delays.

This gives us a fundamental rule:

Greater distance → greater local autonomy.

The Moon can cooperate closely with Earth. Mars requires substantially more independence. More distant infrastructure would require still more.

This is not primarily a design preference. It is imposed by physics.

The Region Must Operate Through Isolation and Failure

Terrestrial applications often treat lost network connectivity as an abnormal condition. Interplanetary systems must expect it.

Communication geometry changes. Relays may fail. Solar activity may interfere with links. Antennas can lose alignment. Systems therefore require **disconnected operation**: local services continue functioning while communication with another region is unavailable, and information is reconciled when connectivity returns.

The same principle applies to maintenance.

Imagine a Martian data centre experiencing hardware failure. Local monitoring detects the fault. An AI system identifies the affected equipment and shifts workloads to healthy machines. A maintenance robot receives instructions and replaces hardware if a spare is available. Earth eventually receives the incident report.

The process resembles modern cloud automation, but with one crucial difference.

On Earth, automation is often convenient.

On Mars, it becomes necessary.

The farther a planetary region moves from immediate human support, the more it must carry its own **digital survival kit**: software repositories, backups, security tools, operational documentation, diagnostic systems and enough local knowledge to recover from failures.

Cybersecurity therefore also becomes increasingly local. A compromised Martian system cannot wait for an emergency security team to physically arrive from Earth. Vulnerabilities must be diagnosed and patched locally wherever possible.

A planetary compute region is therefore not merely a collection of servers. It becomes part of that world's technological self-sufficiency.

Energy, Cooling and Regional Specialization

Although the planets are separated by enormous distances, much of the Solar System shares one major energy source: the Sun.

Solar power already supports satellites and spacecraft and could eventually provide significant energy for off-world computing. But solar intensity decreases with distance from the Sun, while thermal conditions, available materials and communication geometry also vary enormously.

Future infrastructure must therefore ask several questions together:

Where is energy available? Where is computation needed? Where can heat be rejected effectively? Where are useful materials located? Which locations have favourable communication routes?

The best answer may differ according to the workload.

This could eventually produce regional specialisation. Earth would remain the principal human and computing region. The Moon might develop industrial, scientific and archival roles. Mars could eventually support a more autonomous settlement and computing environment. Some asteroids might primarily support resource processing rather than major computation. Other locations could develop specialised purposes according to their physical advantages.

Just as every cloud region today does not need to perform exactly the same role, every planetary region would not need to become identical.

Planetary Data and Governance

Once computing becomes geographically distributed across worlds, familiar governance problems also become larger.

Today, **data sovereignty** describes the principle that information can be subject to the laws of the jurisdiction where it is stored or processed. A multi-planet civilisation eventually raises more difficult questions.

Who governs information generated on the Moon? Should a mature Mars settlement control its own operational data? Should important Earth archives automatically be replicated off-world? Could one government or organisation control a critical communication gateway?

These questions belong partly to later governance discussions, but they affect architecture because technical design determines where control exists.

A completely Earth-controlled system may become increasingly impractical as distant settlements mature. Technical autonomy could eventually contribute to greater political and institutional autonomy as well.

AI and Space Reinforce Each Other

The architecture creates an important feedback loop.

Better AI enables more autonomous infrastructure. Greater autonomy allows machines to operate farther from Earth. Operating farther from Earth creates demand for more local computing. More local computing then supports more capable AI and robotics.

Conceptually:

AI → autonomy → off-world infrastructure → local compute → greater AI capability

This does not mean AI automatically creates a space civilisation. Energy, propulsion, materials, economics and governance remain enormous constraints.

AI instead helps address one unavoidable problem of expansion:

Humans cannot be physically present everywhere civilisation operates.

Autonomous systems can operate equipment, monitor infrastructure and respond to routine failures where immediate human intervention is impossible.

Earth as the Root Region

In the early Solar System Compute Fabric, Earth would remain the root of almost everything. Most hardware would be manufactured here. Most software would be developed here. Most people, universities, research institutions and industrial supply chains would remain here.

Earth would provide the first spacecraft, processors, robots and industrial equipment.

But robust distributed systems gradually reduce unnecessary dependence on their root. As off-world regions mature, they could store more knowledge locally, manufacture more components, maintain their own software and make more operational decisions independently.

Independence does not necessarily mean separation.

A distributed system can contain autonomous regions that continue cooperating as parts of a larger network.

The First Step Is Not Mars

It is tempting to jump directly to the dramatic image of enormous AI data centres operating on Mars. Doing so skips nearly every engineering step required to make them possible.

The progression is much more likely to begin close to Earth:

More capable satellite computing → orbital edge processing → better space communications → autonomous maintenance → lunar communication relays → lunar scientific computing → small lunar compute clusters → robotic industry → local resource utilisation → larger off-world compute regions

Each stage develops technologies needed by the next.

That is why this part of the book is concerned with the **first steps off Earth**, not with pretending that the mature Interplanetary Cloud already exists.

The Moon is the obvious laboratory.

It is close enough for frequent communication with Earth but different enough to force engineers to solve problems involving radiation, cooling, dust, energy, autonomous maintenance, delayed communication and local computing.

The Moon therefore represents much more than humanity's nearest celestial neighbour.

It could become the first place where civilisation learns how to build, power, cool, secure, maintain and operate a genuine compute region beyond Earth—and the first practical step toward a Solar System Compute Fabric.

Chapter 16

Orbital Computing: The First Step Off Earth

Before humanity builds substantial data centres on the Moon or Mars, the first serious expansion of computing beyond Earth is likely to happen much closer to home: **in orbit**.

That transition has already begun in a limited form. Satellites and spacecraft contain onboard computers because they cannot depend on Earth for every decision. They process navigation information, manage sensors, control orientation, compress scientific data and monitor their own systems. As processors improve and AI becomes more efficient, spacecraft can increasingly decide what information matters before transmitting it to Earth.

This is the foundation of **orbital computing**: performing useful computation in space rather than treating every spacecraft as a sensor that simply collects information and sends everything to a terrestrial data centre.

The distinction matters because spacecraft can generate information faster than communication systems can always deliver it.

From Satellite to Edge Data Centre

Consider an Earth-observation satellite continuously photographing the planet. It may produce thousands of images, some containing clouds, empty ocean or information that is not immediately important.

Instead of transmitting everything first and analysing it later, onboard computing can identify cloud-covered images, recognise wildfires, detect unusual weather, identify ships or prioritise valuable scientific observations.

This is **edge computing**—processing information close to where it is generated.

An autonomous vehicle on Earth follows the same principle. It must recognise a pedestrian and brake locally rather than upload camera footage to a distant data centre and wait for instructions. In space, satellites, telescopes and spacecraft similarly benefit from making some decisions close to their sensors.

An orbital computing node could eventually go further. Imagine a larger spacecraft containing substantial compute, storage and high-speed communications. Nearby satellites transfer raw or partially processed information to it. AI performs heavier analysis, information is stored temporarily, and selected results are transmitted to Earth.

At that point, the spacecraft begins to resemble a small **orbital edge data centre**.

Why Put Computing Near the Data?

The advantage is not that processors somehow calculate faster because they are in space. The advantage is that **the data does not need to travel all the way to Earth before becoming useful**.

Suppose a satellite constellation generates one terabyte of observations, but only fifty gigabytes contain information scientists urgently need. Orbital computing could analyse the larger dataset and transmit the important portion first. The complete raw dataset could follow later if required.

As space activity expands, this becomes increasingly valuable. Earth-observation constellations may continuously monitor weather, agriculture, fires, oceans and infrastructure. Telescopes may generate enormous scientific datasets. Lunar industry will eventually produce operational information. Autonomous spacecraft will create continuous telemetry.

Instead of requiring Earth to process everything, some computation moves toward the source.

The same architecture can also combine information from several spacecraft. One satellite observing a cyclone might provide optical images, another radar, another atmospheric measurements and another ocean data. An orbital computing platform could perform **data fusion**, combining these sources into a more useful operational picture before sending the most urgent result to Earth.

Orbital computing therefore does more than save bandwidth. It can turn separate spacecraft into a cooperating computational system.

Communication Is Often the Bottleneck

Space communication depends on transmitting power, antenna size, receiver sensitivity, distance, available frequencies and line of sight.

Traditional **radio-frequency (RF) communication** has supported spaceflight for decades and remains essential. But future missions will produce increasingly large datasets, creating demand for much higher communication capacity.

One promising technology is **optical communication**, which transmits digital information using tightly focused laser light. NASA's Laser Communications Relay Demonstration (LCRD) has tested optical relay technology from geosynchronous orbit, while the ILLUMA-T terminal aboard the International Space Station demonstrated an ISS-to-LCRD optical connection reaching up to approximately 1.2 gigabits per second before the demonstration ended in 2024.

The architecture was essentially:

International Space Station → Laser Link → Orbital Relay → Ground Station

The important lesson is that a spacecraft does not always need a direct communication path to its final terrestrial destination. A relay can sit between the spacecraft and Earth.

Add significant computing and storage to that relay:

Spacecraft → Orbital Compute/Relay Node → Earth

Now the intermediate node does more than forward a signal. It can analyse, prioritise and store information.

As discussed earlier, optical communication increases bandwidth rather than overcoming the speed-of-light limit. Its value here is therefore not lower propagation delay, but the ability to move much larger quantities of data efficiently between spacecraft, orbital nodes and Earth.

From Relay to Orbital Network

A mature orbital architecture could contain different classes of nodes rather than one enormous station.

A sensor satellite might communicate with a nearby orbital edge node. The edge node performs heavier processing and transfers selected information through a high-capacity optical backbone.

Conceptually:

Sensor Satellite → Orbital Edge Node → Optical Backbone → Earth

An **optical backbone** would use laser links between major nodes to provide high-capacity communication across the network.

The concept resembles terrestrial networking. A smartphone does not communicate directly with a data centre on another continent. It connects to nearby infrastructure, which hands the traffic into progressively larger networks.

Space infrastructure could eventually develop a similar hierarchy.

Orbits, Relays and Hops

An orbital data centre is still a spacecraft. It does not simply float because gravity disappears.

Gravity remains substantial in low Earth orbit. A spacecraft stays in orbit because its forward velocity and gravitational acceleration continuously combine so that it falls around Earth rather than directly into it.

The simplified relationship is:

Forward velocity + gravitational acceleration = orbit

The International Space Station travels at roughly 7.7 kilometres per second, or around 28,000 kilometres per hour. Astronauts appear weightless not because Earth's gravity has vanished, but because the station, astronauts and everything inside are falling around Earth together.

Orbit is also not necessarily permanent. In low Earth orbit, traces of Earth's atmosphere produce drag, gradually reducing orbital energy. The ISS therefore periodically requires **reboost manoeuvres**. Higher orbits experience much less atmospheric drag but introduce other trade-offs involving radiation, communication and accessibility.

There is consequently no single best location for orbital computing.

Low Earth Orbit (LEO) offers relatively low communication delay and easier access from Earth but includes atmospheric drag and rapidly moving spacecraft. **Geostationary orbit (GEO)**, approximately 35,786 kilometres above Earth's equator, allows a correctly positioned spacecraft to appear almost stationary over one longitude, making it useful for communications.

Different orbital regions could therefore perform different functions. LEO nodes might process information from nearby satellite constellations. Higher nodes could serve as major communication relays. Other infrastructure could eventually operate closer to the Moon.

This introduces the **hop architecture**:

Satellite → LEO Compute Node → GEO Relay → Earth

and eventually:

Lunar Surface → Lunar Relay → Earth-Moon Gateway → Earth

A **hop** is an intermediate node through which information travels. Importantly, it need not be merely a signal booster. It can contain storage, processing and AI.

If the next link is unavailable, the node stores the information. If data can be compressed or analysed locally, it does so. AI can determine communication priority, monitor equipment and select the next route.

This is where the communication network itself begins to contain **compute and storage**.

Hops Do Not Defeat Physics

Suppose two points are 300,000 kilometres apart. Light takes approximately one second to travel that distance. Placing ten relay stations between them does not reduce the total distance.

Hops therefore cannot turn seconds or minutes of physical light-time into milliseconds.

Their advantages are different. They can regenerate signals, route around failures, store information during outages, support smaller spacecraft and provide local computing services.

This distinction is fundamental to the Interplanetary Cloud.

The architecture works around the limitations imposed by distance. It does not pretend those limitations disappear.

A Laser Backbone Through Space

Optical communication becomes particularly interesting between spacecraft because there are no terrestrial clouds between two nodes already in space.

Atmospheric weather can interfere with laser communication between orbit and the ground, which is one reason multiple optical ground stations can be useful. Between spacecraft, that particular problem largely disappears.

A different difficulty becomes critical: **precision**.

A laser beam is extremely narrow. The transmitting spacecraft must locate and remain accurately pointed toward a receiving terminal while both objects are moving, potentially across enormous distances.

This requires **pointing, acquisition and tracking**.

An intuitive analogy is trying to keep a laser pointed at a moving coin hundreds of kilometres away while both you and the coin are travelling thousands of kilometres per hour.

Future AI and automated network-control systems could continuously predict positions, establish links, allocate bandwidth and redirect traffic as orbital geometry changes.

The Network Must Manage Itself

A terrestrial network engineer can replace equipment or modify configurations from a nearby control centre. An orbital network could contain hundreds or eventually thousands of moving nodes.

Links appear and disappear. Spacecraft enter Earth's shadow. Orbits change. Optical links are interrupted. Relays approach thermal or power limits.

Humans cannot manually optimise every communication decision in real time.

Network orchestration would therefore automatically select routes, allocate bandwidth and move information according to changing conditions. AI could help predict upcoming communication windows, detect unusual behaviour and reprioritise traffic.

The network becomes dynamic rather than static.

Storage makes these nodes even more useful. If a satellite cannot currently reach Earth, it could transfer information to an orbital storage node and continue collecting new observations. The storage node holds the earlier dataset until a high-capacity link becomes available.

This is **store-and-forward communication**.

The same principle becomes far more important when the network eventually stretches between planets.

An Orbital Data Centre Must Survive Space

The phrase *orbital data centre* can create the wrong mental picture. We should not imagine taking an Earth warehouse, putting it inside a giant rocket and leaving it floating in space.

An orbital facility is first a spacecraft.

It requires power, thermal control, navigation, communications, radiation protection, attitude control, collision avoidance and potentially propulsion. Every kilogram matters. Components must survive launch vibration, and failed equipment is much harder to replace than hardware in a terrestrial data centre.

Three challenges become especially important: **radiation, maintenance and debris**.

High-energy particles can interfere with electronics and occasionally change the state of digital components, producing a **single-event upset** such as a flipped bit. Space systems therefore use techniques including radiation-hardened electronics, redundancy and error-correcting memory.

This creates a challenge for AI hardware. Terrestrial accelerators are generally optimised for performance, efficiency and cost, while space-rated processors prioritise reliability in harsh environments. Future orbital AI systems may require semiconductor designs that combine much greater computational performance with improved radiation tolerance.

Maintenance presents another problem. An Earth technician can replace a failed server. Sending astronauts to replace individual orbital processors would be economically unrealistic at large scale.

Orbital facilities would therefore need **modular robotic maintenance**. Workloads move away from a failed module. A robotic arm removes it. A replacement is inserted and automatically tested. Spare components remain available for future failures.

The facility must be designed from the beginning to be repaired by machines.

Earth orbit also contains active satellites, inactive spacecraft, rocket stages and smaller debris. At orbital velocities, even small objects can cause serious damage. Critical equipment therefore requires protection against micrometeoroids and orbital debris, while larger threats must be tracked and avoided.

An orbital computing platform consequently needs collision detection, propulsion and autonomous manoeuvring as parts of its operational architecture.

Power and the Hard Problem of Heat

Solar energy is an obvious source of electricity for orbital computing. There is no atmosphere reducing sunlight before it reaches solar panels, although many Earth orbits periodically pass through the planet's shadow and therefore require batteries or another source of stored energy.

Large AI systems, however, could require vastly more electricity than ordinary satellites.

And almost all of that electrical energy eventually becomes heat.

This creates one of the greatest misconceptions about computing in space. Space may be described as cold, but vacuum does not provide cold air that can be blown through server racks. There is essentially no surrounding atmosphere to carry heat away through convection.

A large orbital computing facility must ultimately reject its waste heat through **thermal radiation**, using radiator surfaces that emit energy into space.

A hypothetical gigawatt-scale AI installation would therefore be an extraordinary engineering project, requiring not only enormous power generation but potentially enormous radiator systems.

This is why the first orbital computing facilities will almost certainly be far smaller and more specialised than today's largest terrestrial AI campuses.

The realistic progression is:

Onboard processing → specialised orbital edge compute → shared compute/relay platforms → larger modular facilities

rather than:

Launch a hyperscale data centre next year.

Modular Growth

Orbital computing may become more practical if infrastructure grows in modules.

One launch could provide power and communications. Another adds computing. Another expands storage. Additional modules provide radiators. Later missions replace obsolete processors.

Robots connect and maintain the system.

This approach also addresses the unusually short life cycle of computing hardware. The entire station should not need to be discarded because its processors become obsolete. Long-lived structural, communication, power and thermal systems could remain while compute modules are periodically upgraded.

Over time, the facility becomes more like infrastructure than a conventional satellite.

From Data Centre to Orbital Service Hub

Once substantial orbital computing exists, its role could expand beyond processing satellite information.

A spacecraft approaching Earth might connect to an orbital hub. Navigation information is exchanged. Diagnostic data is uploaded. AI analyses the spacecraft's condition. Scientific datasets move into orbital storage. Robotic maintenance or replacement components can be scheduled.

At that point, the distinction between **data centre, communications relay and orbital service station** begins to blur.

The computing facility becomes one part of a larger orbital ecosystem.

Humans would not necessarily need to live there. Human habitation requires oxygen, water, food, sanitation, radiation protection and emergency systems. A machine-centred facility could avoid much of this complexity and operate primarily through robotics, with humans visiting only when necessary.

The International Space Station nevertheless provides a valuable engineering preview. It already combines solar power, thermal control, radiators, communications, docking interfaces, navigation, robotic systems, replaceable components, impact protection, resupply and orbit maintenance.

An orbital data centre would use many related systems but organise them around a machine-centred workload rather than a continuously inhabited research laboratory.

The Bridge to the Moon

Once substantial compute and storage can operate reliably in Earth orbit, several later steps become easier.

Satellites can offload processing. Spacecraft gain communication hubs. Lunar missions gain intermediate storage and relays. Robots can access nearby computing services. Planetary backups can begin staging outside Earth. Optical networks gain major routing points. Autonomous spacecraft gain places to exchange navigation and maintenance information.

Most importantly, engineers gain experience operating substantial computing infrastructure in vacuum.

That experience is essential before moving to another world.

The Moon introduces additional difficulties—dust, surface construction, resource extraction and more complex industrial operations—but it also provides something orbital infrastructure cannot:

land and local materials.

Orbital computing is therefore unlikely to become the final destination.

It is the **bridge between the terrestrial cloud and the first true extraterrestrial compute region**.

Once that bridge exists, the next step becomes possible: building a computing and industrial ecosystem on the Moon itself.

PART VI BUILDING BEYOND EARTH

Resources, orbital mechanics, networks, robots and the logistics layer

Closing Part V

Part V moved the Interplanetary Cloud from abstract possibility toward an initial architecture. Orbital computing could provide the first experimental layer, followed eventually by more specialised lunar and planetary infrastructure. But placing machines beyond Earth creates problems that terrestrial cloud architecture rarely encounters at such scale: enormous distances, moving planets, communication delay, launch windows, difficult maintenance and hostile environments. Part VI examines how such an infrastructure could actually be built and operated.

What PART VI Covers

This Part develops the physical and technical architecture of the Interplanetary Cloud. It examines extraterrestrial resources, communication delay, orbital mechanics, launch windows, logistics networks, robotic construction and the changing role of humans as infrastructure moves farther from Earth. It then considers increasingly autonomous systems capable of operating where continuous human control is impossible, solar-powered computing across moving worlds, and the placement of data according to latency: immediate workloads near Earth, selected replicas on the Moon, and delay-tolerant archives farther away. The Part concludes by exploring how data centres could evolve into multi-purpose stations and how remote environments might provide controlled laboratories for testing increasingly autonomous systems.

Chapter 17

What the Solar System Can Give Us

If humanity eventually builds large computing, AI and robotic infrastructure beyond Earth, we cannot indefinitely launch every kilogram of material from Earth's surface.

Chapter 15 introduced the Moon as our first practical training ground for **In-Situ Resource Utilization (ISRU)**—using resources found at the destination instead of transporting everything from Earth. The larger Solar System potentially offers water, metals, silicate rock, carbon-bearing compounds, solar energy and other useful materials.

But finding a resource is only the beginning.

The important questions are:

Where is it? How difficult is it to extract? How much energy does processing require? And is using it actually easier than bringing the same material from Earth?

Earth will remain humanity's overwhelming industrial base for a long time. The objective is not to strip other worlds for resources or suddenly replace terrestrial industry. It is to gradually use local materials for the enormous quantities of water, shielding, structural material, propellant and industrial feedstock that an expanding space infrastructure could require.

Earth: The Industrial Starting Point

Earth begins with an enormous advantage: almost our entire industrial civilisation already exists here.

Semiconductor factories, mines, steelworks, electrical grids, transportation systems, launch facilities and skilled workers are concentrated on one planet. Even if another world contains abundant iron, producing a precision component there is initially far harder than manufacturing it on Earth.

Early extraterrestrial infrastructure will therefore remain highly Earth-dependent. Advanced processors, sensors, precision machinery and sophisticated robotic components will probably continue coming from Earth for a considerable time.

The problem is **mass**.

Launching every structural beam, shielding block, storage tank and construction component from Earth's relatively deep gravitational well would eventually require an enormous transportation system.

ISRU therefore does not mean removing Earth from the supply chain. It means avoiding the unnecessary launch of materials that can eventually be obtained more efficiently elsewhere.

A useful long-term division emerges:

Earth supplies high-complexity technology.

Off-world industry increasingly supplies high-mass basic materials.

The Moon: The First Resource Base

The Moon is the obvious starting point because it is nearby and contains potentially useful resources.

Lunar polar regions contain water ice, although its abundance, accessibility and economic value still require investigation. Lunar regolith also contains oxygen chemically bound inside minerals, together with elements including silicon, aluminium, iron, magnesium, calcium and titanium.

This does not mean lunar dust can simply be turned into a computer.

Processing infrastructure is required. A lunar refinery could heat regolith, separate useful materials and chemically extract oxygen. Other materials could eventually become basic structural products.

The Moon might therefore increasingly provide **water, oxygen, shielding and construction materials**, while Earth continues supplying processors and complex machinery.

One of the most useful lunar resources may also be one of the least glamorous:

regolith itself.

The Moon lacks Earth's protective atmosphere and global magnetic environment. Long-lived infrastructure therefore requires substantial radiation protection. Instead of launching enormous quantities of shielding from Earth, robots could cover habitats and sensitive equipment with local regolith. For computing infrastructure, this could protect electronics; for humans, thicker layers could reduce radiation exposure.

This illustrates an important principle:

The most valuable resource is not necessarily the rarest material. It may be the material that prevents us from transporting enormous mass from Earth.

Mars: A Richer Local Environment

Mars is much farther away but offers a more diverse resource environment.

It contains substantial water ice, including polar and subsurface deposits. Its thin atmosphere consists predominantly of carbon dioxide, giving future industry access to carbon- and oxygen-containing material. Its rocks and soil also contain iron-rich minerals and other potentially useful elements.

The atmosphere itself can become a resource.

NASA's **MOXIE** experiment aboard the Perseverance rover demonstrated this principle by taking carbon dioxide from the Martian atmosphere and producing oxygen electrochemically.

The experiment was small, but the underlying idea is important:

A machine built on Earth can use another planet's environment to manufacture a useful resource.

Water could support people, industry and potentially radiation shielding. Electrolysis can separate it into hydrogen and oxygen, which can contribute to life-support and propellant systems. Water or other fluids could also circulate inside closed-loop thermal systems, although the ultimate waste heat would still need to be rejected.

Mars does possess an atmosphere, unlike the Moon, but its surface pressure is less than one percent of Earth's average sea-level pressure. Conventional terrestrial data-centre cooling therefore cannot simply be copied. Martian infrastructure would require specialised heat exchangers, radiators and possibly subsurface construction.

Mars and Local Manufacturing

Mars becomes particularly interesting if its greater resource diversity eventually allows more complex manufacturing.

Robotic industry could extract water, process atmospheric gases, mine minerals and manufacture structures. Local materials might eventually become roads, landing areas, shielding, storage tanks, habitats, power infrastructure and data-centre enclosures.

Modern semiconductor manufacturing is an entirely different challenge. It requires ultra-pure materials, extraordinarily precise machinery and an immense industrial supply chain. A Martian settlement would therefore not need to manufacture advanced processors before local resources became valuable.

If Mars can manufacture the **heavy and relatively simple components**, Earth can concentrate transportation capacity on the smaller quantities of high-complexity equipment that cannot yet be produced locally.

This same principle will recur throughout an interplanetary economy:

Build locally what is heavy and practical. Import what remains complex and difficult.

Asteroids: Resources Without Large Gravity Wells

Asteroids offer a different opportunity.

Their compositions vary widely. Some are dominated by rocky material, some contain substantial metals, and carbon-rich asteroids may contain hydrated minerals and volatile compounds.

Their major advantage is **weak gravity**.

A small asteroid may have such little gravitational attraction that removing material requires far less energy than launching the same mass from Earth's surface. Some near-Earth asteroids could therefore become useful resource locations for an advanced space economy.

Mining should also avoid transporting unnecessary raw material.

On Earth, we do not normally move an entire mountain to a distant refinery. We extract and process material, then transport the useful portion.

Space industry should eventually follow the same principle.

Robotic spacecraft could survey an asteroid, identify useful material, attach mining equipment, process resources locally and store the valuable products for collection.

There may be little reason for large numbers of humans to live beside such mines.

AI handles immediate decisions. Robots perform physical work. Humans supervise the broader operation from safer locations.

This changes the question from:

Can thousands of humans comfortably live and work here?

to:

Can autonomous machines operate here productively and safely?

That dramatically expands the industrial map of the Solar System.

Mercury: Energy-Rich but Difficult

Mercury is interesting because it lies much closer to the Sun, where solar intensity is several times greater than at Earth.

At first glance, that sounds ideal for energy-hungry computing and industry.

But Mercury is not simply a giant solar-powered data-centre site. Its sunlit surface reaches extreme temperatures, while nighttime conditions are dramatically colder. It has almost no substantial atmosphere for conventional cooling, and thermal management would be extremely difficult.

Reaching Mercury is also more complicated than simply falling toward the Sun.

A spacecraft launched from Earth already carries Earth's substantial sideways orbital velocity around the Sun. Reaching Mercury requires removing a large amount of orbital energy, which is why missions often use carefully designed trajectories and **gravity assists**.

This introduces a concept that becomes increasingly important to space logistics: **delta-v**.

Delta-v measures the change in velocity required to perform a manoeuvre or journey. In practical spaceflight, physical distance alone does not tell us how difficult a destination is to reach.

The important question is not merely:

How many kilometres away is it?

It is:

How much velocity must we change to get there, land, leave and deliver the cargo?

Mercury also demonstrates how unintuitive planetary resources can be. Despite being the planet closest to the Sun, permanently shadowed polar regions appear to contain water ice.

Location, geometry and local conditions matter as much as a planet's overall reputation.

Venus: Resource-Rich but Extremely Hostile

Venus presents almost the opposite problem.

It is approximately Earth-sized and relatively nearby, but its surface environment is extraordinarily hostile. Its atmosphere is overwhelmingly carbon dioxide, surface pressure is roughly ninety times Earth's sea-level pressure, and temperatures are around 465°C.

That makes conventional surface data centres, mines and human settlements highly unattractive.

Venus is nevertheless not resource-free. Its atmosphere contains enormous quantities of carbon dioxide and other compounds, while conditions higher in the atmosphere are considerably less hostile than those on the surface.

Future robotic systems might investigate atmospheric resources or specialised high-altitude platforms.

For the architecture explored in this book, however, Venus is better considered a **long-term robotic frontier** than an early industrial target.

The Moon, near-Earth asteroids and Mars offer more practical initial steps.

Robots Change the Map

This leads to one of the central arguments of this book.

If we ask:

Where can humans economically live and work?

much of the Solar System looks unusable.

But if we ask:

Where can AI-controlled machines operate?

the map changes dramatically.

A robot does not need oxygen, food or conventional living space. Machines can be designed for radiation, vacuum or temperature conditions that humans cannot tolerate. If communication with Earth disappears temporarily, sufficiently autonomous systems can continue performing assigned tasks.

AI provides decision-making.

Robotics provides physical action.

Local resources provide material.

Together, they form the industrial foundation required for large off-world infrastructure.

Building Machines Beyond Earth

Serious extraterrestrial mining and construction equipment could eventually become enormous. Launching every completed machine inside a single rocket may be impractical.

The solution is likely to be **modular construction**.

Earth sends motors, electronics, precision mechanisms and other high-complexity components. Structural sections arrive separately or are increasingly manufactured locally. Robotic assembly systems connect them at the destination.

Imagine sending the **brain and precision components** of a mining machine from Earth while manufacturing much of its heavy frame from lunar material.

That could substantially reduce launch mass.

Eventually robots could assemble other robots—not necessarily manufacture every advanced electronic component but build complete machines from a mixture of imported precision parts and locally produced structures.

Coordinating such operations requires another layer: the **AI foreman**.

A future construction site may contain no human standing beside the machinery. One robot surveys terrain. Another excavates. Another transports material. A refinery processes it. Inspection drones examine structures. Assembly robots install components. Maintenance robots replace failed modules.

The AI foreman coordinates the fleet according to human-defined objectives, safety limits, energy availability, resource budgets and maintenance schedules.

If a machine fails, work is redistributed. If solar power drops, non-essential operations slow. If a transport route becomes unavailable, another is selected.

The industrial system gradually becomes capable of operating for longer periods without direct human intervention.

Automated Transport Completes the Chain

Mining a resource is useful only if that resource can reach somewhere it is needed.

Launch, landing and cargo systems must therefore become increasingly autonomous.

Modern spacecraft already use automated guidance, navigation and control, while reusable launch vehicles have demonstrated autonomous landing. Future cargo networks could extend the same principle much further.

A robotic facility prepares a shipment. Autonomous machines load it. The spacecraft performs checks and launches. Navigation software controls its trajectory. It reaches an orbital logistics hub, where robots unload the cargo. Another vehicle eventually carries it to its next destination.

Humans supervise the overall system rather than manually controlling every movement.

The model begins to resemble **container shipping across orbital routes**.

Space Needs a Logistics Layer

Computing cannot expand across the Solar System without physical logistics.

Servers fail. Solar panels degrade. Robots break. Radiators are damaged. Propellant is consumed. Humans need supplies. Mining facilities need replacement components.

A **logistics layer** is the transportation and supply network that keeps those systems functioning.

Modern cloud computing can appear almost weightless because users interact with software through screens. Behind that software lies an enormous physical system of semiconductor factories, fibre cables, electricity networks, cooling equipment, transport and technicians.

The Interplanetary Cloud would require an even larger physical foundation.

This is why the future of AI cannot be separated from **energy, mining, manufacturing, robotics and transportation**.

The digital world ultimately depends on matter.

From Global Supply Chain to Solar Supply Chain

A mature space economy would probably not make every world completely self-sufficient.

Specialisation is more efficient.

Earth might remain the principal source of advanced technology and human expertise. The Moon could increasingly provide water, oxygen, shielding and basic construction materials. Asteroids might eventually supply particular minerals or volatiles. Mars would develop local resources primarily to support its own infrastructure. Orbital facilities could assemble or service components where manufacturing in space becomes advantageous.

Resources would flow toward the locations where they provide the greatest value.

This is not fundamentally different from terrestrial trade.

A modern computer may contain minerals extracted in one country, processors fabricated in another, memory manufactured elsewhere, software developed globally and final assembly performed somewhere else.

A future lunar structure might similarly contain lunar aluminium, Earth-manufactured processors and components assembled in orbit. A Martian data centre might use locally manufactured shielding around computing modules originally built on Earth.

What appears to be a "Martian data centre" could therefore be the product of an entire **Solar System supply chain**.

The Interplanetary Cloud becomes a network of specialised regions rather than isolated colonies.

Expansion Without Repeating Earth's Mistakes

Access to extraterrestrial resources does not remove environmental responsibility.

The objective should not be to transfer Earth's destructive extraction practices into space.

Not every lunar crater needs to become a mine. Not every asteroid needs to be dismantled. Scientifically important locations may require protection, and future discoveries could create entirely new conservation priorities.

Resource extraction should be justified by genuine need and governed responsibly.

The advantage of expanding beyond Earth is not that environmental responsibility disappears. It is that civilisation gains access to vastly more physical space and material without requiring every expansion of computing and industry to occur inside Earth's biosphere.

Earth is not simply another industrial location.

It is the only world currently known to support billions of humans and an extraordinarily complex natural biosphere.

Protecting it should remain a priority.

A Distributed System, Not a Dyson Sphere

The architecture described here should also be distinguished from the famous **Dyson sphere** concept.

A Dyson sphere broadly describes hypothetical structures designed to capture a very large fraction of a star's energy.

That is not necessary for the future proposed in this book.

We do not need to surround the Sun or capture all of its energy. We need progressively better ways to position computing, storage, communications, energy generation and industry where each makes practical sense.

Solar arrays around Earth. Orbital compute nodes. Lunar industry. Communication relays. Martian infrastructure. Robotic resource operations. Perhaps much later, specialised facilities closer to the Sun.

The architecture is **distributed**, not one giant megastructure.

That is much closer to how technological systems actually evolve.

Resources Are Only Half the Problem

The Solar System potentially provides much of the physical foundation needed for expansion: water for life and industry, oxygen locked inside ice and minerals, metals and silicates for construction, carbon-bearing compounds for chemistry, enormous solar energy and vast physical space for robotic infrastructure.

But knowing that a resource exists does not tell us how to reach it.

Mars is moving.

The Moon is moving.

Earth is moving.

Asteroids follow their own orbits.

And the spacecraft carrying our equipment is moving as well.

Space contains no fixed roads connecting stationary destinations.

Before we can build the logistics layer of the Interplanetary Cloud, we therefore need to understand the physics that determines where those roads actually are.

This also leads to an important correction to the original idea behind this book. What I initially thought of as **"centrifugal interplanetary gravity"** is better understood through gravitational fields, orbital velocity, reference frames, transfer orbits and Lagrange points.

These concepts explain something remarkable:

How can we launch a spacecraft toward a planet moving millions of kilometres through space and arrange for both to arrive at the same place at the same time?

That is the subject of the next chapter:

Orbital Mechanics Without the Mathematics.

Chapter 18

Distance Changes Everything

The previous chapter examined what the Solar System could provide: water, minerals, energy and physical space for an expanding technological civilisation. But resources are useful only if we can reach them, communicate with them and move information and equipment between them.

On Earth, distance has become almost invisible to digital technology. Messages cross oceans in fractions of a second. Cloud systems replicate information between countries. Engineers remotely manage infrastructure thousands of kilometres away.

Space changes that assumption completely.

The Moon is about 384,400 kilometres from Earth on average. Mars is not at one fixed distance because Earth and Mars both orbit the Sun; their separation varies by tens to hundreds of millions of kilometres. Other planets continuously change their distance from Earth as well.

The future Interplanetary Cloud therefore cannot simply extend terrestrial networking outward and expect it to behave the same way.

Distance becomes part of the architecture.

The Speed of Light Becomes Visible

Light in vacuum travels at approximately **299,792 kilometres per second**. Radio waves, microwaves, infrared and laser light are all forms of electromagnetic radiation, so none provides a way around this fundamental limit.

The Moon makes the limit noticeable. A signal takes roughly 1.28 seconds to travel the average Earth-Moon distance one way. Even under ideal conditions, a request and response therefore require more than 2.5 seconds simply because of physics.

Mars changes the experience completely.

Depending on the relative positions of Earth and Mars, one-way communication generally takes roughly 3 to 22 minutes. If the delay is twenty minutes one way, Earth must wait approximately forty minutes for the earliest possible response before adding any processing time.

This delay is **latency**.

On Earth, engineers often try to reduce network latency to milliseconds. Between planets, millions of kilometres make substantial latency unavoidable.

More powerful lasers cannot solve this problem.

Optical communication can greatly increase **bandwidth**—the amount of information transmitted during a period of time—but the first bit of information still cannot arrive faster than light.

This distinction is fundamental:

Bandwidth determines how much data can travel.

Latency determines how long it takes to arrive.

Future communication technology can dramatically improve the first. It cannot eliminate the second.

That physical limitation actually strengthens the case for distributed computing.

Distance Forces Local Intelligence

Imagine a Martian cooling system detecting dangerous overheating.

If it takes twenty minutes to ask Earth what to do and another twenty minutes for the instruction to return, the equipment may already have failed.

A mining robot cannot stop for forty minutes every time it encounters an unexpected obstacle. A spacecraft approaching Mars cannot depend on a human on Earth making every last-second navigation decision.

Distance therefore creates a direct architectural relationship:

Distance → latency → autonomy → local AI → local computing

The farther civilisation expands, the more intelligence must travel with it.

This is already visible in Mars exploration. Engineers do not drive rovers as though they were remote-controlled cars. Mission teams establish objectives and transmit instructions, while onboard systems perform parts of the operation locally. NASA's Perseverance rover, for example, has autonomous navigation capabilities that allow it to analyse terrain and make driving decisions without Earth directing every movement.

The principle is:

Earth defines objectives. Local intelligence handles immediate execution.

A future Martian industrial region containing thousands of robots, power systems, habitats, communication networks and data centres would require this principle on a much larger scale.

Earth could remain an important source of knowledge, software and strategic direction.

It could not micromanage Mars.

Distance Changes Data Replication

The same physical limitation affects data.

On Earth, cloud systems can replicate information between nearby facilities extremely quickly. Across planets, a remote copy will inevitably lag behind its source.

Suppose Earth maintains an important civilisation archive on Mars. When information changes on Earth, Mars cannot know about the update milliseconds later. The information must first cross interplanetary space.

The architecture therefore requires **asynchronous replication**: changes are accepted at one location and transferred to another later without forcing the original system to wait for confirmation from the distant copy.

For some workloads, that delay would be unacceptable. A financial transaction platform cannot casually tolerate twenty-minute inconsistencies.

But a civilisation archive has a different purpose.

If its job is to preserve human knowledge against catastrophic loss, a copy that is twenty minutes, an hour or even a day behind Earth may still provide extraordinary resilience.

Terrestrial IT already uses a related concept: the **Recovery Point Objective (RPO)**, which defines how much recent data an organisation is prepared to lose after a major failure.

Interplanetary systems could extend that logic.

Earth's operational systems remain highly current. Lunar replicas may follow seconds behind. Martian archives may follow minutes or longer behind. Different information receives different replication policies according to its purpose.

The design principle is:

Do not fight latency where physics makes latency unavoidable. Design around it.

Hops Improve Reliability, Not Light-Time

Intermediate relay nodes become valuable here.

Imagine:

Earth → Earth-Orbit Node → Deep-Space Node → Mars-Orbit Node → Mars

At first, it might seem that dividing the journey into smaller hops would reduce latency. It does not.

If Earth and Mars are 300 million kilometres apart, dividing those 300 million kilometres into ten segments does not shorten the total physical distance. Each relay may actually introduce a small additional processing delay.

So why build the hops?

Because **latency is not the only communication problem**.

Connections can disappear. Spacecraft moves behind planets. Equipment fails. Communication resources are shared. Planetary geometry changes. Solar interference can make some periods difficult.

An intermediate node can receive information, verify it and store it. If the next link is unavailable, the information waits safely. When communication becomes possible again, the node forwards it.

The architecture becomes:

Send → Store safely → Confirm → Forward when possible

This is **store-and-forward networking**, and it is far better suited to a network spanning moving planets than assuming one continuous end-to-end connection will always exist.

The Network Can Become Part of the Storage System

Store-and-forward also creates an interesting possibility.

Suppose Earth sends an encrypted incremental backup toward Mars in millions of chunks. An orbital node receives and verifies some of them. Those chunks are temporarily safe there. They later move onward while the node receives newer information.

During the transfer, pieces of the backup may exist throughout the network:

Earth: complete primary copy
Earth Orbit: recent outgoing chunks
Deep-Space Node: verified in-transit chunks

Mars Orbit: destination staging data
Mars: durable secondary archive

The network itself has become part of the storage architecture.

This does not make the Martian copy instantaneous. It provides **incremental resilience**. If a link fails, verified information that has already reached a reliable intermediate node may not need to restart its entire journey from Earth.

The Internet of Space Must Tolerate Disruption

Today's Internet generally assumes that a usable route to the destination exists now.

An interplanetary network cannot safely make that assumption.

This is why **Delay/Disruption-Tolerant Networking (DTN)** is important for space communication. DTN is designed for networks where connections may have long delays or disappear temporarily. Instead of discarding information because the next destination cannot currently be reached, a node can retain it until forwarding becomes possible.

A simple analogy is freight transportation.

A conventional Internet packet behaves somewhat like a passenger making a mostly continuous journey.

Interplanetary data behaves more like freight:

Travel → Warehouse → Wait → Next transport connection → Destination

For digital information, the warehouse is storage.

The network is designed around interruption rather than treating interruption as an exceptional failure.

Sometimes the Best Transmission Is No Transmission

Distance also makes **caching** extremely valuable.

Imagine a Mars engineer requesting a technical manual stored only on Earth. Waiting perhaps twenty minutes for the request to arrive and another twenty minutes for the document to return would make routine work painfully slow.

The solution is not faster-than-light communication.

It is to keep commonly required information on Mars.

A cache stores information close to where it is likely to be needed. Martian infrastructure could maintain local copies of software packages, engineering manuals, medical knowledge, maps, scientific databases, repair documentation and AI models.

The same principle applies to the Moon, orbital hubs and spacecraft.

As distance increases, **data placement becomes as important as data transmission**.

The best interplanetary network request may be the one that never needs to cross interplanetary space.

AI Must Move Closer to Its Users

This becomes critical for AI.

Imagine a Mars settlement whose primary operational AI runs entirely in an Earth data centre.

A technician asks:

"Why has oxygen pressure fallen in Habitat 4?"

Waiting tens of minutes for an answer would be unacceptable.

Mars therefore needs local AI capable of analysing its own infrastructure. Larger models could periodically receive updates from Earth, while specialised models operate even closer to individual machines.

A mining robot might contain its own navigation intelligence. A habitat might run a specialised safety model. A planetary data centre might host larger models shared across the settlement.

This creates **hierarchical AI**:

Device AI → Facility AI → Planetary AI → Interplanetary services

Not every problem needs the largest AI model available to civilisation.

Immediate problems need intelligence close enough to respond immediately.

Software Must Learn Patience

Distance also changes application design.

Modern software frequently assumes that one service can call another and receive a response almost immediately.

On Earth:

Request → response in milliseconds or seconds

Across planets:

Request → minutes of travel → processing → minutes of return travel

A conventional synchronous application may become unusable under those conditions.

Interplanetary software therefore needs greater reliance on **messages, queues, local processing and delayed synchronisation**.

A message queue stores a request until the receiving system is available to process it. Instead of Earth asking Mars a question and freezing while it waits, Earth sends a message and continues working. Mars processes the request when it arrives and eventually returns the result.

Software becomes designed around patience.

This is not merely slower cloud computing.

It is a different architectural assumption.

Security Must Also Become Local

Suppose Earth discovers a serious vulnerability affecting software used throughout the Solar System.

The Moon can receive a patch within seconds. Mars receives it minutes later, assuming communication is available. More distant infrastructure could wait much longer.

A planetary region therefore cannot depend on Earth for every security decision.

It requires local authentication, monitoring, recovery software, backups, security policies and the ability to detect abnormal behaviour. AI may assist with identifying threats and responding when Earth cannot be contacted.

This is a form of **technical operational sovereignty**.

A distant region must contain enough capability to protect and operate itself even when communication with its parent world is unavailable.

Human Control Changes Too

Humans naturally prefer immediate control over important machines.

Across planets, that becomes impossible.

Instead of:

Human → command → machine obeys immediately

the relationship increasingly becomes:

Human → objective + rules + limits → autonomous system executes → system reports results

Humans establish goals, boundaries and safety requirements. Local systems make many immediate operational decisions.

That demands more than intelligent AI.

It requires predictable behaviour, extensive testing, strong governance and systems capable of failing safely when humans cannot intervene immediately.

Autonomy therefore increases naturally with distance:

Earth: frequent human intervention
Earth orbit: significant automation with rapid support
Moon: substantial autonomy with Earth seconds away
Mars: strong autonomy with Earth minutes away
Outer Solar System: very high autonomy with communication potentially taking hours

This also creates a sensible development ladder.

Build and test autonomous systems on Earth. Operate them in orbit. Expand them on the Moon. Learn from failures. Then deploy increasingly capable descendants farther into the Solar System.

Logistical Distance Matters Too

Communication delay is only one form of distance.

There is also **logistical distance**.

A lunar data centre may be only 1.3 light-seconds from Earth, but a failed component could still take days or longer to replace physically. Cargo to Mars takes months.

That changes infrastructure design.

A terrestrial data centre can often assume replacement equipment will arrive relatively quickly. A Martian facility cannot.

Its maintenance model must move from:

Fail → order replacement

toward:

Predict → repair → reuse → manufacture locally where practical

Systems need spare components, modular hardware, robotic repair and eventually some local manufacturing.

AI can monitor equipment continuously and use **predictive maintenance** to identify components likely to fail before they actually do.

Distance therefore encourages infrastructure to become less disposable.

Knowledge must also arrive before emergencies occur. A Martian facility needs local technical documentation, diagnostic AI, repair robots, training systems, spare parts and qualified local personnel where humans are present.

A highly skilled engineer on Earth cannot simply travel to Mars tomorrow because a cooling pump has failed.

Distance does not remove human work.

It changes what work needs to exist locally.

Distance Forces Decentralization

The larger lesson is that the Solar System cannot operate as one centrally controlled data centre.

Earth cannot remain the processor for every robot, database and operational decision.

Instead:

Earth manages Earth.
The Moon increasingly manages the Moon.
Mars manages Mars.
Orbital systems manage local traffic.
Spacecraft manage their immediate environments.

Regions exchange software, knowledge and selected information, but each retains enough intelligence and infrastructure to continue operating independently.

In computing terms, civilisation moves from a largely **centralised architecture** toward a **distributed architecture** in which responsibility is spread across cooperating systems.

The speed of light effectively forces that transition.

Distance Also Creates Resilience

Enormous separation is not purely a disadvantage.

Distance also provides isolation.

A terrestrial disaster cannot physically destroy a Martian data centre. A major Martian infrastructure failure does not automatically disable Earth. A cyberattack affecting one planetary region might potentially be isolated from another. Important information can exist in physically independent locations.

Distance creates latency.

But distance also creates **failure-domain separation**.

When properly designed, that separation can create resilience.

Important information moves between them according to priority, replication policy and available communication paths. No single network connection needs to remain permanently available. No single planet needs to contain every copy. And no human needs to make every operational decision.

This is no longer simply cloud computing placed in space.

It is a **Solar System-scale distributed system**.

But there is another complication.

Earth, Mars, the Moon and every relay between them are moving. Earth travels around the Sun at roughly 30 kilometres per second. Mars follows a different orbit at a different speed. The Moon moves around Earth. Every relay placed in space must follow an orbit or trajectory of its own.

There is no straight, stationary highway from Earth to Mars where we can simply position data centres every few million kilometres.

The entire network is moving.

Understanding how spacecraft, relays and future infrastructure remain in useful positions therefore requires us to temporarily leave networking and examine the physics underneath it.

That brings us to **orbital mechanics—and why nothing in space simply stays where we put it.**

Chapter 19

You Do Not Aim at Mars

One of the easiest mistakes when imagining interplanetary infrastructure is to think of the Solar System as a map with fixed destinations. Earth is here, Mars is there, so we launch a spacecraft toward Mars. But Mars will not still be there when the spacecraft arrives. Earth and Mars are both orbiting the Sun, at different speeds and on different-sized orbits. A spacecraft leaving Earth is already travelling around the Sun with Earth before its engines even fire. Interplanetary navigation is therefore not simply a question of pointing toward another planet. **You aim for where the planet will be when your spacecraft gets there.** This principle is essential to the Interplanetary Cloud. Future cargo vehicles, mining robots, communication relays and data-centre modules will all operate inside a Solar System where almost everything is moving. There can be no permanent straight highway between Earth and Mars. Instead, future infrastructure must understand orbital motion well enough to predict where planets, spacecraft and network nodes will be days, months and years ahead.

Earth Is Already Moving

Although we feel stationary while standing on Earth, our planet travels around the Sun at roughly **30 kilometres per second**, or more than 100,000 kilometres per hour. Mars travels at roughly **24 kilometres per second** on average. Earth completes one orbit in about 365 days, while Mars requires about **687 Earth days**. This difference means the relative positions of the planets continuously change. Sometimes Earth and Mars are comparatively close; at other times they are on widely separated parts of their orbits. It is also why efficient opportunities to launch conventional missions from Earth toward Mars occur only about every 26 months. A launch window is a period when the positions and motions of two worlds make a particular journey practical. For a future space economy, launch windows could become something like shipping schedules today, except the timetable is partly determined by gravity and planetary motion.

A Spacecraft Does Not Start from Zero

Another important point is that a rocket sitting on Earth is not stationary relative to the Sun. It is already travelling around the Sun with Earth. When the spacecraft launches, its engines modify the motion it already has. This is why travelling to another planet is better understood as changing from one orbit to another rather than driving through space in a straight line. Earth follows one path around the Sun. Mars follows another. The spacecraft must enter a trajectory that carries it from the neighbourhood of Earth's orbit toward the neighbourhood of Mars's orbit. This path is called a transfer orbit. One simplified example is the Hohmann transfer, an energy-efficient trajectory between two approximately circular orbits. The spacecraft receives an initial change in velocity that stretches its solar orbit outward from Earth toward Mars. It then spends much of the journey coasting under gravity. If the launch was timed correctly, Mars reaches the spacecraft's arrival region at approximately the same time.

Conceptually:

Earth at departure → transfer orbit → Mars at its future position

The spacecraft is therefore not chasing the Mars we can see when it launches. It is arranging an interception months into the future.

Distance Is Not the Same as Difficulty

This also explains why kilometres alone are a poor measure of how difficult a space journey will be. Engineers use another important concept called **delta-v**. Delta-v means the amount by which a spacecraft must change its velocity to complete a manoeuvre or journey. A destination can be relatively close yet require a large velocity change. Another can be farther away but accessible through a more favourable trajectory. Propellant requirements depend heavily on those changes in velocity, not merely on kilometres travelled. This gives us two different measurements that matter enormously to the Interplanetary Cloud: Distance determines much of our communication delay. Delta-v helps determine much of our transportation difficulty. That is why the Moon, Mars, Mercury, asteroids and orbital stations cannot simply be ranked according to their distance from Earth.

Gravity Does Not Disappear in Space

This brings us to an idea I originally described while developing this book as **"centrifugal interplanetary gravity."** My underlying question was valid: if we put communication and data-centre nodes into space, how do we keep them synchronised with planets that are continuously moving? But the scientific explanation is different. There is no single centrifugal gravitational corridor connecting the planets. What we need to understand is orbital mechanics—the interaction of gravity, velocity and trajectories. Gravity does not suddenly disappear when we enter space. The International Space Station, for example, remains strongly affected by Earth's gravity. It stays in orbit because it is moving sideways fast enough that, as gravity pulls it toward Earth, Earth's curved surface continually falls away beneath it. An orbit can therefore be understood as continuous free fall around a body. Centrifugal force, meanwhile, is an apparent outward effect that becomes useful when describing motion from a rotating frame of reference. It is not a new force that can hold a line of data centres between Earth and Mars.

The correction actually makes our architecture more interesting. We do not need the network to remain stationary. We need the network to understand motion.

Why We Cannot Put Fixed Hops Between Earth and Mars

Earlier chapters introduced the idea of intermediate data-centre hops:

Earth → Hop 1 → Hop 2 → Hop 3 → Mars

As a networking diagram, this remains useful. Physically, however, the nodes cannot simply be parked at fixed points along a straight line. Earth is orbiting the Sun. Mars is orbiting the Sun at a different rate. Each relay placed between them would also follow an orbit or trajectory determined by its velocity and the gravitational forces acting on it. **The neat straight line would therefore continually change.** A future interplanetary network needs to look more like a *moving mesh. A mesh network is a network containing multiple possible routes between nodes rather than depending on one permanent path.* At one time, Earth might have a good communication path through Relay A. Hours or days later, Relay B might provide a better route. Another relay may temporarily lose useful line of sight. Mars-orbiting infrastructure may become available through a different path. Instead of expecting the network topology to remain fixed, we design it to change continuously.

We Already Use Dynamic Routing on Earth

The idea of a network changing its route is not new. Modern computer networks already use **dynamic routing protocols**—systems that allow routers to learn about available paths and adjust when the network changes.

One example is **OSPF (Open Shortest Path First)**, a routing protocol widely used inside terrestrial networks. Imagine that information can normally travel from City A through City B to City C. If the connection through City B fails, routers can learn that the path is no longer available and calculate another route through the network.

The important idea is not the details of OSPF. It is that the route is **not permanently written into the network**.

Earth-based routing, however, usually operates under an assumption that becomes much harder to maintain across the Solar System: usable communication paths are expected to exist reasonably continuously. A broken terrestrial link is treated mainly as a failure that the network should route around.

In space, a missing link may not be a failure at all.

A relay may be functioning perfectly but temporarily unable to communicate with another node because the planets, spacecraft and relay stations are moving. The network may know that the connection is unavailable now—and also know that it will become available later.

That changes the routing question.

A terrestrial router often asks:

Which available path should carry this information now?

An interplanetary node may sometimes need to ask:

Where should I store this information now so that it can continue its journey when the next communication path becomes available?

Dynamic routing therefore does not disappear beyond Earth. It evolves. The network must consider not only **where its nodes are**, but **where they will be and when they will next be able to communicate**.

The Network Can Predict Its Future

Fortunately, planetary motion is not random. We can calculate where planets will be. We can predict spacecraft trajectories. We can estimate when communication links will become available or disappear. This creates an unusual form of networking.

A terrestrial router asks:

Where should I send this information now?

An interplanetary router may need to ask:

Where should I store this information now so that it can be forwarded when the next connection becomes available? Suppose Relay A can communicate with Earth now but cannot currently reach Relay B. Orbital calculations indicate that a suitable connection will become available later. Relay A stores the data, waits for that communication window, and then forwards it. This is where Delay/Disruption-Tolerant Networking (DTN) becomes particularly useful. Instead of assuming that every end-to-end path exists continuously, the network can work around predictable periods of disconnection. The Solar System network becomes partly a network and partly a timetable.

From a Moving Mesh to a Distributed Data Pipeline

A future interplanetary network would probably not operate as a simple chain in which every piece of information follows exactly the same route from Earth to Mars. Because planets, spacecraft and relay stations are constantly moving, the most useful communication path would change with time. A better architecture would therefore be a **dynamic mesh**, meaning a network containing many possible routes between nodes. A relay that has a good connection with Earth today may have a better connection with another relay tomorrow. The network would continuously calculate which links are available, which are likely to become available and where information should wait until its journey can continue. This develops the earlier idea that an interplanetary network would become partly a network and partly a timetable.

Distance explains why many intermediate nodes could eventually become useful, although there is no reason to assume that thousands of evenly spaced relays would be required between Earth and Mars. Earth and Mars can be tens of millions to hundreds of millions of kilometres apart. If hypothetical relays were spaced only 100,000 kilometres apart, a 225-million-kilometre separation would require roughly 2,250 links. If reliable communication links could span one million kilometres, the same route would require about 225 links, while links spanning ten million kilometres would reduce that to roughly 23. The correct spacing would therefore depend on communication technology, power, pointing accuracy, storage capacity, orbital geometry, maintenance requirements and cost. The objective would not be to create the largest possible number of relays. It would be to create enough overlapping paths that the network could continue operating when individual links became unavailable.

The important point is that these hops would not make information travel faster than light. Dividing a 225-million-kilometre journey into 225 smaller sections still leaves approximately 225 million kilometres for the signal to cross. Their purpose would instead be resilience. Each important relay could receive information, verify it, temporarily store it and forward it when another suitable route became available. This is an extension of the **store-and-forward** architecture already discussed earlier in this book, where intermediate nodes become part of both the communication system and the storage system.

Large transfers would also not need to travel as one enormous object. Imagine Earth needs to send a one-petabyte incremental backup toward Mars. The update could be divided into perhaps millions of smaller encrypted data units, depending on the future protocol and network design. Each unit could carry information identifying the archive, its version, its position within the larger transfer and an integrity value allowing the destination to confirm that the data arrived correctly. Different pieces could then travel through different parts of the mesh. One group might move through an Earth-orbit facility and Relay A, another through Relay B, and another through a lunar or deep-space node. They would not necessarily arrive in the same order in which they left Earth. Mars would collect the verified pieces, identify what was still missing and reconstruct the complete update once the required information had arrived.

The idea is similar in principle to how modern networks divide information into manageable units, but an interplanetary system would have to tolerate interruptions lasting minutes, hours or potentially much longer. For particularly important information, the system could even create additional recovery fragments so that losing one route or relay would not necessarily require retransmitting the entire missing section from Earth. Temporary copies would exist throughout the network while the transfer was progressing. Earth would retain the authoritative source, intermediate nodes would hold selected verified pieces, Mars-orbit infrastructure might accumulate an almost complete update, and the Mars data centre would eventually commit the completed version.

The result would therefore not be a fixed line of thousands of data centres stretching between two planets. It would be a **moving, predictive mesh of communication, storage and computing nodes**. Routing software—and eventually AI—could decide where each piece of information should go according to current and predicted orbital geometry, available bandwidth, energy, storage capacity, congestion, equipment health and the urgency of the data. A node might effectively ask: *Which available neighbour gives this information the best chance of reaching its destination within the required recovery window?* As the planets move, the answer changes. The Interplanetary Cloud would continuously reorganise its routes around the Solar System rather than expecting the Solar System to remain still for the network.

Lagrange Points Provide Useful Regions

If our interplanetary network needs communication, storage and computing nodes in space, an obvious question appears: **where do we put them?** We cannot simply place a data-centre relay halfway between Earth and Mars and expect it to remain there. Earth moves around the Sun, Mars moves around the Sun at a different speed, and the relay itself must also follow an orbit. This is why **Lagrange points** become interesting to the Interplanetary Cloud.

A Lagrange point is a useful region created by the orbital relationship between two large bodies, such as the Sun and Earth or Earth and the Moon. There are five commonly identified points, called **L1, L2, L3, L4 and L5**. At these locations, the gravitational influence of the two larger bodies and the orbital motion of a much smaller object combine in a way that allows the smaller object to maintain a useful position relative to them with comparatively manageable orbital control.

A Simple Way to Picture It

Imagine two people running around a large room. One is a tall adult running quickly, while the other is a shorter child moving at a different speed and along a different path. If the adult throws a ball directly at where the child is standing now, the ball will almost certainly miss because both people continue moving while the ball is in the air. The only reliable way to make the pass is to predict where the child will be and throw the ball toward that future meeting point rather than the child's current position.

Space works in much the same way. Earth, the Moon, Mars and every relay station are continuously moving through space at different speeds and in different orbits. Lagrange points are not places where everything stops moving. Instead, they are predictable geometric regions created by orbital motion and gravity where spacecraft can maintain particularly useful relationships with the larger bodies. For the Interplanetary Cloud, these regions become valuable because they provide stable, predictable locations where communication, navigation or storage nodes can operate as part of our moving network of hops.

The five points also have different geometries. **L1** lies broadly between the two larger bodies. **L2** lies beyond the smaller body on the opposite side. **L3** lies on the far side of the larger body, while **L4 and L5** form triangular arrangements roughly 60 degrees ahead of and behind the smaller body's orbit. L4 and L5 can be dynamically more stable in suitable two-body systems, whereas spacecraft operating around L1 or L2 generally require periodic **station-keeping**—small propulsion manoeuvres used to keep them near their intended trajectories.

We already make practical use of these regions. The **James Webb Space Telescope** operates in an orbit around the Sun-Earth L2 region, roughly 1.5 million kilometres from Earth. It does not sit motionless at L2. It follows an orbit around that region while maintaining a useful relationship with Earth and the Sun. This geometry helps keep the Sun, Earth and Moon broadly on the same side of the telescope, which is particularly valuable for its observations and thermal design.

Now return to our network of moving data-centre hops. We should not assume that every future relay would use a Lagrange point—most would not, and different nodes could occupy many kinds of planetary, lunar and heliocentric orbits. But **some Lagrange regions could become valuable anchor locations within the larger moving mesh**. A suitable node might provide communications, temporary storage, navigation support, computation or logistics services while maintaining a predictable relationship with Earth, the Moon or another part of the network.

For example, instead of imagining a permanent straight route:

Earth → Hop 1 → Hop 2 → Hop 3 → Mars

we can imagine a much more flexible architecture:

Planetary nodes + orbital nodes + selected Lagrange-region nodes + deep-space nodes → dynamic interplanetary mesh

The routing system would know where these nodes are now and where they are expected to be later. A Lagrange-region node could therefore become one useful junction among many possible routes. Data might reach it, be verified and stored, and then wait for another suitable connection to become available. At another time, orbital geometry might make a completely different relay the better route.

This is why Lagrange points matter to the Interplanetary Cloud. **They do not create shortcuts through space, reduce the speed-of-light delay or provide magical gravity-free locations. They provide potentially useful regions around which parts of a moving communications, computing and logistics network could be organised.** Combined with ordinary planetary orbits, heliocentric relays and other spacecraft trajectories, they give our future network something it desperately needs: useful and predictable places within a Solar System where almost everything is moving.

A Hop Is Really an Orbiting Data Node

We can now improve the definition of a future interplanetary hop. It is not a stationary signal tower floating between planets. It is an orbiting or travelling infrastructure node whose future position is predictable. A substantial node might contain communications equipment, computing, storage, AI, solar generation, batteries, navigation systems and robotic maintenance. When a connection is available, it receives data. It verifies that data. It stores it if necessary. It processes selected information locally. When the next appropriate link becomes available, it transmits the information onward. The data therefore travels through a network whose nodes themselves travel through space. This is very different from today's fibre network, where routers are normally fixed inside buildings.

AI Could Become the Network Traffic Controller

Now imagine this architecture at scale. Thousands of satellites operate around Earth. Lunar relays orbit the Moon. Cargo spacecraft travel between destinations. Scientific probes travel farther outward. Mars has surface and orbital infrastructure. Some communication links use radio. Others use optical lasers. Nodes have different amounts of energy, bandwidth, storage and processing capacity. No human team could continuously optimise every possible path manually. AI could become the traffic controller of the interplanetary network. A future network-management AI could know the predicted position of each node, available energy, communication windows, storage capacity and priority of different information. It might send emergency navigation information through one path while allowing a large scientific archive to wait several hours for a higher-capacity optical connection. This is dynamic routing—selecting and changing network routes according to conditions. The same AI could predict future congestion, equipment failures or periods of poor connectivity.

The network becomes increasingly autonomous because, once again, distance makes continuous human control impractical.

Chapter 20

Moving Through a Moving Solar System

Communication Hops and Spacecraft Stops Are Different

There is one important distinction. A communication signal passing through a relay does not mean a spacecraft should physically visit that relay. The most useful route for data and the most efficient route for a spacecraft may be completely different. A communication relay forwards information. A logistics hub allows spacecraft to exchange cargo, receive maintenance, refuel or dock. Some future infrastructure could combine both roles, particularly around Earth, the Moon and Mars, but we should not assume that every communication hop becomes a physical rest stop. Spacecraft must perform a rendezvous to meet another spacecraft or station. A rendezvous requires the two objects to reach approximately the same position with compatible velocities. Changing trajectory to visit an inconveniently located station consumes propellant. Therefore, rather than evenly spaced "service stations" between Earth and Mars, major logistics hubs would more likely emerge in orbital locations that naturally support transportation routes.

Rest Stops in Space Could Still Become Important

This does not eliminate the original rest-stop idea; it makes it more realistic. A large orbital infrastructure hub could eventually combine computing, communications, storage, spacecraft diagnostics, robotic maintenance, cargo transfer and perhaps protected human accommodation. A spacecraft might dock there because the location is already useful to its journey. Crew members could rest. Scientific data could be transferred. Software and navigation information could be updated. Robots could inspect the spacecraft. Replacement components could be installed. Cargo could move between vehicles. Eventually some hubs could contain propellant depots—facilities storing propellant for visiting spacecraft. Instead of a motorway service station, the better analogy is a port. Ships do not stop every hundred kilometres across the ocean. They travel between strategically located ports that provide fuel, repairs, supplies and cargo transfer. Future spacecraft may do something similar.

The Moon Could Help Supply Those Ports

The Moon could eventually become one source of material for these orbital hubs. If lunar industry develops, water, shielding materials and selected structural products might be supplied from the Moon rather than lifted from Earth. The economics and supply-chain implications of that possibility are examined in the next chapter.

Cargo and Humans Need Different Routes

A future logistics system also does not need to move everything at the same speed. Human crews may value shorter journeys because long-duration spaceflight increases exposure to radiation, isolation and life-support risks. Bulk cargo may tolerate much longer journeys. A replacement computer needed urgently on Mars might use an expensive faster transfer. Thousands of tonnes of non-urgent structural material could follow slower, more energy-efficient trajectories. We already make the same distinction on Earth. Urgent goods travel by aircraft. Bulk commodities cross oceans by ship. Interplanetary logistics will probably develop similar service levels. The Solar System may eventually have its own equivalent of express freight and bulk shipping.

Robots May Arrive First—but Humans May Still Be Needed

Robotic missions will probably reach many future construction sites before human settlers do. That does not mean, however, that we already possess robots capable of autonomously building an industrial civilisation on Mars.

Today's robots can perform increasingly sophisticated tasks, but the kind of machines imagined later in this book—large mining systems, autonomous construction fleets, self-maintaining factories and machines capable of helping build other machines—would represent a much more advanced stage of technology.

Developing such systems may itself be difficult on Earth. Smaller versions can be built and tested here, allowing engineers to improve autonomous navigation, construction, mining, maintenance and safety. But experimenting with extremely large autonomous machines near populated regions could create legitimate concerns about land use, environmental impact, regulation and what happens when experimental systems fail. Earth is a living environment shared by billions of people, not an unrestricted industrial laboratory.

This creates a more gradual path into space.

In the nearer term, robotic missions can deliver equipment, survey sites, deploy communication systems and perform limited construction before people arrive. Infrastructure can also expand around Earth, while the Moon—much closer than Mars—can become an important testing ground for robotic construction, local resource use, energy systems, habitats and eventually specialised computing facilities.

Mars would be a much greater challenge. Early robotic missions could prepare sites and position equipment, but the first substantial Martian settlements may still require humans. Engineers and technicians could supervise machines, solve unexpected problems, repair failures and help establish the industrial ecosystem that later generations of robots would depend upon.

Over decades—and perhaps more than a century—the balance could gradually change. Humans and smaller robots might initially build and operate the first facilities. Those facilities could support larger machines. Local materials could reduce dependence on Earth. Robotic maintenance could become more capable. Eventually, sufficiently mature extraterrestrial industrial regions might support large autonomous machine fleets that would be extremely difficult to develop or operate at comparable scale on Earth.

The progression might therefore look like this:

Small robotic systems tested on Earth → orbital and lunar infrastructure → robotic preparation of Martian sites → early human settlements → human-supervised industrial expansion → increasingly capable autonomous machines → large robotic industrial ecosystems → increasingly self-maintaining extraterrestrial infrastructure

The important principle is not that robots must replace humans before settlement begins. It is that **humans and machines can build the conditions that allow later generations of machines to operate with progressively greater independence.**

For destinations that are especially hostile to human life, such as Mercury—and perhaps some future Venusian operations—the balance could be different. There, increasingly capable robotic systems may eventually perform most or all physical operations without permanent human settlement.

Automation Must Extend to Launch and Landing

This future depends on more than intelligent mining robots. Transportation itself must become increasingly autonomous. We are already moving in that direction. Modern spacecraft routinely perform significant portions of navigation, guidance and control automatically, and reusable rocket systems have demonstrated automated landing capabilities.

Future cargo transport could extend automation across the entire chain:

robotic factory → autonomous loading → launch → orbital navigation → rendezvous → unloading → onward transport → planetary landing Humans might supervise thousands of such movements without manually piloting every spacecraft. AI could coordinate launch windows, orbital traffic, cargo priorities and equipment availability. In effect, the Solar System develops an automated logistics platform.

Nothing Is Truly "In Sync"

This finally answers the original question of how our future infrastructure can remain in sync with moving planets. It does not remain synchronised by travelling at one common speed. Earth has its orbit. Mars has its orbit. The Moon has another. Communication relays have theirs. Logistics hubs have theirs. Spacecraft follow transfer trajectories between them. The system works because those movements are predictable. Software knows where each important node is expected to be. It knows when communication links will exist. It knows when rendezvous opportunities occur. It knows how long information should take to arrive. AI continually updates those predictions using real observations. The network is synchronised through knowledge of motion, not through physical stillness.

From a Map to a Timetable

This requires us to change how we imagine the Solar System.

Do not picture:

Earth — Station — Station — Station — Mars Picture multiple moving circles, trajectories and nodes whose relationships continually change. A useful analogy is an enormous railway timetable. A station does not need every train permanently connected to every other station. The system works because the location and timing of trains are known. Interplanetary infrastructure will be vastly more complicated, but the principle is similar. A data packet may wait for its next connection. A cargo spacecraft may wait for a launch window. A lunar shipment may wait for an orbital rendezvous. A Mars mission may depart only when planetary geometry becomes favourable. Gravity establishes much of the timetable. Computers calculate it. AI continuously optimises it. Robots execute the physical work. Humans establish the objectives and governance. This is how the Interplanetary Cloud stops being a collection of imaginary stationary data centres and becomes a physically plausible moving infrastructure network. And it leads naturally to the next problem.

Knowing where infrastructure needs to go is only the first problem. The next is keeping it supplied. Servers, robots, replacement parts, water, structural materials and people cannot travel as information does. They require a physical transportation and supply network. Once the Solar System becomes a moving network of destinations, civilisation therefore needs something more than orbital mechanics: it needs interplanetary logistics.

Chapter 21

The Interplanetary Logistics Network

The previous chapters established that a space-based AI civilisation would require more than a network for moving information. Radio signals, laser communications, software, AI models and replicated data can travel electronically, but servers, processors, robots, solar panels, radiators, water, minerals, replacement parts and people cannot. They have mass, and moving mass through space requires spacecraft, energy, time and carefully planned trajectories.

This physical supply chain is the **Interplanetary Logistics Network**.

Today's Internet may appear almost weightless to its users, but every digital service depends on an enormous physical supply chain. Semiconductor factories manufacture processors. Mines provide raw materials. Ships and aircraft transport equipment. Trucks deliver servers. Technicians replace failed components. Moving data centres beyond Earth would not eliminate this physical system; it would extend it across hundreds of thousands and eventually millions of kilometres.

If the Interplanetary Cloud becomes the digital nervous system of a future space civilisation, its logistics network will be the system that keeps its physical body alive.

From Global Supply Chains to Solar System Supply Chains

A terrestrial data centre can order a replacement server and potentially receive it within days. A lunar data centre cannot call a local courier. A Martian facility faces a much greater problem: depending on planetary positions and mission design, physical cargo can take months to arrive from Earth.

This changes infrastructure design fundamentally. A Mars facility cannot remain dependent on an emergency shipment from Earth every time a pump, robotic joint or structural component fails. It would need spare parts, predictive maintenance, repair capability and eventually the ability to manufacture some components locally.

The supply chain could therefore gradually expand from Earth manufacturing through launch facilities, orbital hubs, lunar infrastructure and interplanetary transportation to Mars and other destinations. Not every shipment would pass through every stage. Sometimes a direct route would be more efficient. The important change is that space logistics could gradually become a **network of routes, warehouses and transportation hubs**, rather than Earth independently launching everything directly to its final destination.

Earth Exports Complexity; Space Increasingly Supplies Mass

For a long time, Earth would remain the manufacturing centre of this network. Advanced processors illustrate why.

A modern AI accelerator may be physically small, but producing it requires semiconductor fabrication plants, extremely pure materials, specialised chemicals, precision machinery and a vast international supply chain. Reproducing that entire industrial ecosystem on the Moon merely to avoid transporting processors from Earth would initially make little sense.

Early extraterrestrial industry could instead concentrate on materials and products that are heavy but comparatively simpler to produce. Structural components, radiation shielding, water, oxygen, storage tanks and selected construction materials might eventually be produced using local resources, while processors, advanced sensors and other precision equipment continue arriving from Earth.

This creates a useful principle for the early space economy:

Earth exports complexity. Space increasingly supplies mass.

The boundary would not remain fixed. As extraterrestrial industry becomes more capable, increasingly sophisticated products could eventually be manufactured locally.

Does This Kilogram Really Need to Come from Earth?

Everything manufactured on Earth must first escape Earth's gravity before it can join the space economy. This is one reason reusable launch systems matter so much. Reusing major parts of a launch vehicle can improve the economics of access to space, but it does not make launch free. Rockets still require propellant, maintenance, infrastructure and enormous amounts of energy, while payload mass remains limited.

Future engineers would therefore repeatedly confront a simple question:

Does this kilogram really need to come from Earth?

For an advanced processor, the answer may remain yes for a long time. For thousands of tonnes of radiation shielding or structural material, the answer may eventually be no.

The same economic logic would determine how quickly cargo travels. Urgent equipment might justify a faster and more expensive journey, while structural materials could follow slower, more energy-efficient trajectories. We already make similar choices on Earth: urgent goods travel by aircraft while bulk commodities cross oceans by ship.

Interplanetary logistics would therefore optimise not simply for speed, but for **mass, urgency, energy, cost and destination**.

Orbital Warehouses and Space Ports

As the previous chapter explained, a future transportation network would not consist of stationary motorway-style rest stops positioned along a straight line between planets. Spacecraft, planets and stations are all moving, and changing trajectory to rendezvous with an inconveniently positioned station requires energy.

The more useful analogy is a **port**.

Major logistics facilities could develop at strategically useful orbital locations around Earth, the Moon and eventually Mars. Selected Lagrange-region or other orbital facilities might provide specialised services where their geometry makes them useful.

An orbital facility could also operate as a warehouse. Imagine a group of servers, robotic components and other equipment being launched together from Earth. Some cargo remains in Earth orbit, some travels toward the

Moon and some waits for a later interplanetary transport opportunity. Robots sort and transfer the cargo between spacecraft.

This is conceptually similar to terrestrial logistics centres. A parcel travelling internationally may pass through several warehouses and transportation systems before reaching its destination. In space, however, the transfers would be governed by orbital mechanics and transportation windows rather than roads and shipping lanes.

Propellant Depots Could Change Space Transportation

One particularly important type of orbital warehouse could be a **propellant depot**—a facility storing materials spacecraft can use for propulsion away from Earth's surface.

Propellant should not be confused with petrol. Rockets already use several different propulsion systems and propellants. Some chemical rockets use combinations such as liquid hydrogen and liquid oxygen, while other systems use different fuels and oxidisers. Future spacecraft could use a wider range of technologies, including electric propulsion and potentially nuclear-powered propulsion.

This introduces an important distinction: **an energy source and a propellant are not necessarily the same thing**. A future nuclear reactor, for example, could provide energy for a propulsion system while the spacecraft still uses another material that is heated or accelerated and expelled to produce thrust.

Carrying propellant creates an important transportation problem because propellant itself has mass and launching that mass requires additional propellant. If spacecraft could replenish some propulsion resources away from Earth, future mission design could become more flexible.

The Moon becomes particularly interesting here. If lunar water can eventually be extracted economically, it could support human life and industrial processes, while hydrogen and oxygen derived from it could potentially support some forms of spacecraft propulsion.

A future supply chain might therefore develop from **lunar resources to processing facilities, orbital depots and spacecraft**, rather than requiring every useful kilogram to be lifted from Earth's surface.

The exact resources required would depend on the propulsion technologies used at the time. Chemical, electric and future nuclear propulsion systems would not necessarily need the same infrastructure. A future orbital depot may therefore resemble less a petrol station and more a **multi-purpose energy and propulsion supply hub**.

None of this becomes economical merely because resources exist on the Moon. Mining, processing, storing and transporting them would require substantial infrastructure. But if lunar industry already exists to support settlements, manufacturing, energy systems and data centres, supplying the wider transportation network could become another use for the same industrial base.

Logistics Hubs Could Also Become Compute Hubs

This is where the physical and digital networks begin to overlap.

A major orbital logistics facility would already require substantial computing. It must track spacecraft, manage docking, monitor cargo, control robots, calculate trajectories, maintain inventories, communicate with other facilities, monitor power systems and detect equipment problems.

Once significant computing, communications and storage exist there, the same facility could also become one of the nodes of the Interplanetary Cloud.

A strategically important platform might therefore combine three roles:

Logistics hub + communication relay + compute/storage node

This does not mean every data centre should become a space station. Specialisation would remain useful. But at important locations, sharing power, communications, robotic maintenance and other infrastructure could make combining some functions more economical.

AI Becomes the Logistics Coordinator

A mature interplanetary supply chain could eventually become too complicated for people to schedule manually. Thousands of cargo items might move between Earth, orbital facilities, the Moon and Mars. Each has a destination, mass and priority. Spacecraft have limited capacity. Launch opportunities change. Equipment fails. Energy and communication conditions vary.

AI could help coordinate this system in much the same way that optimisation software already assists terrestrial supply chains, but with much longer delays and much higher consequences for mistakes.

A logistics system could decide which cargo should receive priority, which items can wait for slower transportation, which shipments can be combined and where supplies should be stored before the next stage of their journey. Humans would establish objectives, priorities and safety constraints while increasingly capable software manages the enormous number of operational decisions within those boundaries.

Predictive Maintenance Becomes Part of Logistics

The best emergency shipment is the one that never becomes an emergency.

This makes **predictive maintenance** particularly valuable. Predictive maintenance uses sensor data and operational patterns to estimate when equipment may fail before the failure actually occurs.

Imagine an AI system monitoring thousands of coolant pumps inside a lunar compute region. It detects a gradual change in the vibration of one pump and estimates that the component may fail within several months. The logistics system checks the local inventory and discovers that no replacement is available. Instead of waiting for the pump to fail, it adds the replacement component to an upcoming shipment.

For Mars, where resupply may take months, this ability becomes even more important.

Maintenance and logistics therefore become connected:

Monitor → predict → order → transport → replace

The supply chain begins responding to failures before they happen.

Spare Parts Become Distributed Data

Future extraterrestrial facilities may not need to store a physical replacement for every component. For parts that can be manufactured locally, they could sometimes store the **design** instead.

Imagine that a machine fails. AI identifies the damaged component. A digital engineering library contains its specifications. An automated manufacturing system retrieves the design and produces a replacement using available local material. A robot then installs it.

Three-dimensional printing provides an early example of converting digital designs into physical objects, although future extraterrestrial manufacturing would involve many processes beyond today's 3D printers.

Some future spare-parts warehouses could therefore contain fewer physical objects and far more manufacturing information.

Whenever practical, transport information instead of mass.

But there are important limits. Manufacturing a structural bracket is very different from manufacturing a cutting-edge AI processor. Modern semiconductor fabrication requires extraordinary precision, specialised equipment, controlled environments and complex supply chains. Having silicon available on Mars would not mean a settlement could manufacture advanced processors.

Local manufacturing would therefore develop in layers. Relatively simple structural materials and components could come first, followed by increasingly sophisticated mechanical systems. Advanced electronics may remain imported from Earth much longer.

The objective is not instant self-sufficiency. It is the gradual reduction of dependence on the longest and most fragile parts of the supply chain.

Standardization Makes the Network Scalable

This system would become extremely difficult to maintain if every organisation used completely different components.

Imagine a lunar facility containing dozens of types of robotic arms, each requiring different motors, connectors, batteries and mounting systems. The spare-parts inventory would become enormous.

Space infrastructure therefore creates strong pressure for **standardisation**. Future systems could use compatible battery modules, robotic connectors, cargo containers, computing modules, docking interfaces and replacement parts. A maintenance robot should ideally be able to remove a failed module and install a compatible replacement without requiring the entire surrounding system to be redesigned.

Terrestrial shipping provides an important lesson. One of the greatest advances in global logistics was not simply a faster ship—it was the standard shipping container. Standard containers allowed ships, trains, trucks, cranes and ports to interact with the same physical unit.

Space logistics could eventually develop an equivalent idea. A standard cargo module could leave an Earth factory, be loaded onto a launch vehicle, transferred robotically at an orbital warehouse, carried toward the Moon or another destination and unloaded by another robotic system without people repeatedly repacking its contents.

Standardisation is how a collection of individual machines begins to become an infrastructure system.

Robots Will Operate Much of the Physical Network

The farther logistics expands from Earth, the less practical it becomes to depend upon large numbers of human workers for routine cargo handling.

Orbital warehouses could use robotic systems to identify, sort and transfer cargo. Lunar terminals could become increasingly automated. Mars would require even greater autonomy because communication delays make continuous remote control difficult.

A cargo spacecraft might arrive autonomously. Robotic systems secure it, verify its manifest, unload standard cargo modules and update the inventory. Urgent components move immediately toward their destination while other supplies enter storage. Outbound cargo is loaded for the spacecraft's next journey.

Humans would still travel through this network, particularly during the long period when extraterrestrial infrastructure requires engineers, scientists, technicians and other specialists. But transporting people is fundamentally different from transporting machines. Humans require breathable air, water, food, temperature control, radiation protection, sanitation, medical support and reliable emergency systems.

As discussed earlier, this does not necessarily mean that the same people must permanently settle on Mars. Early human operations could use rotating teams undertaking long assignments before eventually handing their responsibilities to replacement crews and returning to Earth. **The infrastructure and positions can remain even when the people rotate.**

Water would consequently become an especially useful logistics commodity. The same broad resource can support human life, industrial processes and some forms of radiation shielding, while products derived from water could potentially support certain propulsion systems. This is one reason accessible extraterrestrial water could become strategically important to a future space economy.

Recycling Becomes Economically Necessary

A failed component on Earth may sometimes be cheaper to discard than repair. That mentality becomes much more expensive when the replacement is millions of kilometres away.

A damaged metal component still contains processed material. A failed solar array contains materials that required considerable effort to transport or manufacture. A retired server still contains structural components, electronics and other potentially useful resources.

Extraterrestrial industry would therefore have strong economic reasons to repair, reuse, refurbish and recycle wherever practical. Space may force future infrastructure toward a more **circular economy**, where waste is treated as a resource because replacing everything from Earth is simply too expensive.

Interestingly, technologies developed for this purpose could also become useful on Earth. Better automated repair, modular design, recycling and resource efficiency are not problems unique to Mars or the Moon.

Logistics Creates New Fields of Work

This reconnects with the employment question discussed earlier in the book. AI may automate some existing work, but building infrastructure across the Solar System would also require forms of engineering and operations that barely exist today.

People would be needed to design orbital supply chains, autonomous cargo systems, robotic maintenance platforms, modular infrastructure, space-rated computing systems, resource-processing equipment, life-support systems and software capable of operating across enormous communication delays.

Many of those people might remain on Earth. A future Mars infrastructure engineer would not necessarily live permanently on Mars. Humans could design, analyse and supervise systems from Earth while rotating specialist crews and increasingly autonomous machines perform the physical work where necessary.

AI therefore does not merely change existing jobs. It can also change what civilisation becomes capable of attempting.

The Logistics Network Makes the Interplanetary Cloud Physical

The Interplanetary Cloud cannot be built from software alone.

AI requires processors. Processors require machines. Machines require materials. Materials require extraction and manufacturing. Equipment requires transportation and maintenance. Human operations require life support. All of it requires energy.

The architecture therefore forms a reinforcing chain:

Resources → manufacturing → transport → infrastructure → computing → AI → automation → more infrastructure

Each layer helps make the next possible.

This is why the future of AI eventually becomes inseparable from the future of physical engineering. We began this book with people worrying that AI might replace human work on Earth. We are now reaching the other side of that argument. If civilisation chooses to expand responsibly, AI could help us undertake engineering projects that are too distant, dangerous or large for conventional human labour alone.

But one major link is still missing.

A rocket can deliver twenty components of a large mining machine to the Moon. A logistics system can know where every component is. An AI system can hold the complete assembly instructions. None of that produces a functioning machine until something physically connects those pieces together.

Sending human construction crews everywhere would recreate many of the limitations that automation is intended to overcome.

The next stage is therefore not simply better transportation.

It is a new kind of workforce:

robots capable of building machines larger than themselves—and eventually helping to build the machines that build the next generation of infrastructure.

That is where the next chapter begins.

Chapter 22

Robots Build the Machines That Build the Future

The Interplanetary Logistics Network can deliver equipment to orbit, the Moon and eventually Mars, but delivery solves only half the problem. Someone must unload the cargo, assemble the machinery, connect power and communications, inspect the result, repair failures and replace equipment as it ages. Sending large human construction teams to every location would be enormously expensive and, in many environments, unnecessarily dangerous. The workforce that makes large-scale extraterrestrial infrastructure possible will therefore need to be heavily robotic. This does not mean humans disappear from the system. Humans may live permanently on the Moon and Mars, and some large orbital or deep-space hubs could eventually contain human accommodation where there is a genuine operational or transportation reason. But the relationship changes. Humans design, supervise and govern the infrastructure; increasingly capable robots perform much of the dangerous, repetitive and physically demanding work.

Do Not Launch the Giant Machine—Launch the Ability to Build It

One of the original ideas behind this book was that the machinery required for extraterrestrial industry could eventually become too large to conveniently launch as complete units. Imagine a mining machine hundreds of tonnes in mass, an enormous radiator structure for an AI data centre, or kilometre-scale solar infrastructure. Trying to fit the finished system inside a rocket would make little sense. Instead, we can launch modules. A module is a standardised component designed to connect with other components. Structural sections can arrive separately from motors, electronics, computing systems and robotic arms. Once delivered, autonomous construction machines assemble them at the destination. We already use a primitive version of this principle in space. The International Space Station was assembled from many major components delivered over numerous missions. Future infrastructure could extend the idea while shifting much of the assembly work from astronauts to robots.

The progression could become:

Earth manufactures components → rockets transport modules → robots unload them → AI verifies inventory → robots assemble the

machine → AI tests it → machine enters service. Eventually, even some of the modules may no longer need to come from Earth.

Earth Sends the Brain; the Moon Builds More of the Body

Chapter 17 introduced a useful division between **complexity and mass**. Advanced processors, sensors and precision components may remain easier to manufacture on Earth for a long time. Large structural components could increasingly be produced from extraterrestrial resources. Imagine a future lunar excavator. Earth supplies its processors, cameras, specialised motors and control electronics. Lunar industry produces part of its frame, shielding and other heavy structural components. Robotic assembly systems combine them. Later generations could contain a larger proportion of locally manufactured parts. This creates a gradual path toward self-sufficiency without requiring the unrealistic assumption that the first lunar settlement immediately recreates Earth's entire industrial economy. The same principle could eventually operate on Mars.

Robots Will Need More Than Remote Control

Early extraterrestrial robots can receive substantial instructions from Earth, particularly on the Moon where communication delay is only a few seconds for a round trip. Mars is different. Depending on planetary positions, one-way communication takes minutes. A human operator therefore cannot control a Martian construction robot as though operating an excavator through a video game. The robot needs autonomy—the ability to perform assigned tasks and make limited local decisions without requesting human instructions for every movement.

Humans might specify:

Construct the solar-array foundation inside this approved area according to Design 7. The local AI then divides that objective into tasks. Survey robots map the site. Excavators prepare the terrain. Transport robots move material. Assembly robots install components. Inspection robots verify the construction. Earth supervises the objective. The machines handle much of the execution.

Self-Correcting Does Not Mean Uncontrolled

A future robot should also be able to recognise when something has gone wrong. Suppose an assembly robot attempts to install a component but detects that the connector is misaligned. Instead of repeatedly applying force until something breaks, it stops, inspects the connection, compares the observed geometry with its engineering model and attempts an approved corrective procedure. This is a form of fault detection and recovery: identifying an abnormal condition and taking predefined or bounded actions to restore normal operation. More capable AI could improve diagnosis, but autonomy should not mean unlimited freedom to rewrite machinery or invent dangerous repairs. There should be layers of authority. A robot might independently replace a standard failed module. A more unusual repair could require approval from a local supervisory AI. A major structural modification could require human engineering approval from Earth. The architecture therefore combines autonomy with permission boundaries.

Robots Will Repair Other Robots

Once thousands of machines operate away from Earth, another requirement becomes unavoidable: the robots themselves will fail. Motors wear. Bearings degrade. Radiation affects electronics. Dust enters mechanisms. Sensors malfunction. Micrometeoroids damage exposed structures. A system in which every robotic failure requires an astronaut defeat much of the purpose of robotics. Future infrastructure therefore needs maintenance robots specifically designed to inspect and repair other machines. One robot may remove a failed wheel assembly from another. A robotic arm may replace a compute module. Inspection drones may examine solar arrays and radiators. A specialised machine may clean lunar dust from equipment. The failed component can then be repaired, recycled or returned to a manufacturing facility. The robotic workforce becomes partly responsible for maintaining itself.

But What Repairs the Maintenance Robot?

This produces an interesting engineering recursion. Robot A repairs the data centre. Robot B repairs Robot A.

What repairs Robot B?

The answer is not an infinite chain of specialised robots. It is **redundancy and standardisation**. Several robots should be capable of overlapping tasks. Components should use common interfaces wherever practical. Spare robotic units should exist. Critical equipment should not depend on one unique machine. If Maintenance Robot A fails, Robot B can perform enough of A's functions to recover it. If both fail, another unit can be transferred from a nearby facility.

This is essentially the same principle already used in resilient computing:

Do not design a critical service around a single point of failure. The idea applies equally well to physical machines.

The Same Principle Applies to the Hops

Now consider the interplanetary compute and communication hops introduced earlier. A future hop might contain storage, AI compute, radio and optical communications, navigation equipment, solar arrays, batteries, radiators, propulsion and robotic maintenance systems.

What happens if something breaks?

For an ordinary component failure, onboard systems could first attempt recovery. Software reroutes workloads away from failed hardware. Robotic systems inspect the physical equipment. A standard module is replaced if a spare exists. Earth can provide instructions where necessary, although the instructions may take seconds or minutes to arrive depending on the location. But this still assumes the hop remains operational enough to diagnose itself.

What if it does not?

What If the Entire Hop Goes Offline?

Suppose a significant impact or major electrical failure disables the entire node. Its main computers stop responding. Communications disappear. Its maintenance robot is damaged as well. No amount of remote instruction from Earth can repair a machine that cannot receive the instruction. This is where the solution moves from component redundancy to infrastructure redundancy. In cloud engineering, we do not normally protect an important service by saying, "This server must never fail." We assume it eventually will.

Instead, another server is capable of taking over. The same philosophy can extend into space.

Zonal Redundancy in Space

Imagine an important orbital region containing three infrastructure nodes:

Zone A — Hop A

Zone B — Hop B

Zone C — Hop C

The word zone here is conceptual: a deliberately separated group or location designed so that one failure does not automatically disable the others. It is analogous to the availability-zone idea used in terrestrial cloud architecture, although actual orbital placement would have to follow orbital mechanics rather than conventional geography. The three nodes do not necessarily need identical orbits, but they should be arranged so that one local event is unlikely to destroy all of them. Normally, traffic may be distributed across the available nodes. Then Hop A suddenly disappears. The network detects the loss of its heartbeat. A heartbeat is a small recurring signal that tells another system, I am still operating. When several expected heartbeats are missed, the network declares Hop A unavailable. Traffic moves through B and C. The network remains operational. This is failover: transferring a service from failed infrastructure to functioning infrastructure. The interplanetary network therefore does not depend on repairing Hop A before communication can continue.

Redundancy Must Be Physical, Not Merely Digital

There is an important difference between running two applications on one spacecraft and having two independent spacecraft. If one impact destroys the spacecraft, both applications disappear. True resilience therefore requires physical separation. Power should be independent. Communications should have alternative paths. Critical data should exist elsewhere. Maintenance capability should not live exclusively inside the equipment it is expected to repair.

This gives us a powerful rule for future space infrastructure:

Never place the only recovery mechanism inside the failure domain it is supposed to recover. If Hop A contains the only robot capable of repairing Hop A, a catastrophic failure defeats the design. A neighbouring node, planetary base or logistics hub should retain additional recovery capability.

A Repair Mission Can Then Begin

Failover keeps the service alive but losing a billion-dollar or trillion-dollar future infrastructure node should not mean abandoning it immediately. Once Hop A becomes unavailable, the network determines its last known orbit and condition. Telescopes, radar or nearby spacecraft attempt to locate it. If the structure remains intact enough to recover, a robotic repair vehicle can be dispatched from a nearby service location. This might be Earth orbit for nearby infrastructure, lunar orbit for Earth-Moon systems, or eventually Mars orbit for Martian infrastructure. The repair vehicle carries replacement modules and specialised tools. It performs a rendezvous, matching its trajectory with the damaged platform. Inspection robots examine the structure. If recovery is possible, damaged modules are isolated and replaced. Power is restored. Communications are tested. Navigation systems determine the platform's exact trajectory. The node rejoins the network. This is effectively a space-based field service operation.

We Cannot Simply "Put It Back" in Orbit

There is one refinement to the original idea. If an impact damages a hop, we cannot necessarily repair it and simply place it back into its previous orbit as though putting a server back into a rack. The collision may have changed its velocity or trajectory. Engineers—or autonomous navigation software—would first determine its new orbit. If sufficient propulsion remains, carefully calculated manoeuvres could restore it to an appropriate operational orbit. If that requires too much propellant, the better option might be to move it into another useful orbit. If it is unsafe or irreparable, the remaining components could be salvaged where practical and the object moved toward an appropriate disposal trajectory. Space infrastructure must therefore include orbital recovery, not merely hardware repair.

What About an Asteroid Strike?

The word *asteroid* should be used carefully here. For infrastructure around Earth, routine collision risks are more likely to involve orbital debris and micrometeoroids than a large asteroid. Even small objects can be dangerous because orbital collision velocities can be enormous. Large objects can often be tracked, allowing spacecraft to perform collision-avoidance manoeuvres where necessary. Very small particles are harder to avoid and may require shielding and damage-tolerant design.

Future infrastructure would therefore use several layers of defence:

tracking → avoidance → shielding → compartmentalisation → redundancy → failover → repair

Compartmentalisation means dividing a system so damage to one section does not automatically destroy the rest. No individual technique makes a platform indestructible. Together they make catastrophic service loss less likely.

What If an Entire Zone Is Lost?

Now imagine something worse. A major debris event or other local hazard makes one orbital region temporarily unsafe. This is why zonal redundancy should eventually extend beyond individual neighbouring platforms. A service might exist across multiple orbital locations.

For example:

Earth Orbital Zone A

Earth Orbital Zone B

Earth-Moon Zone Critical information exists in more than one location. Communication has alternative routes. If Zone A becomes unavailable, traffic shifts toward Zone B.

This resembles terrestrial cloud architecture:

Machine redundancy protects against machine failure. Zonal redundancy protects against local infrastructure failure. Regional redundancy protects against larger geographic failure.

The Interplanetary Cloud extends the same logic:

Component → Hop → Orbital Zone → Planetary Region → Interplanetary Region

Resilience scales with civilisation.

Humans May Live on Some Hops

Some larger hubs could also contain people. This is physically possible even though the planets move because, as Chapter 20 explained, the hub itself follows an orbit. The International Space Station already demonstrates that humans can live for months aboard an orbiting platform. A future orbital logistics and computing hub might contain crew quarters for engineers, scientists, spacecraft crews or emergency maintenance personnel. But humans should be present only where their presence provides enough value to justify the enormous life-support requirements. A human-rated hub needs breathable atmosphere, pressure control, water, food, sanitation, radiation protection, temperature regulation, medical facilities and emergency escape capability. That makes it much more complicated than a purely robotic data centre. Therefore, I would expect a mixture: Uncrewed hops perform routine communication, storage and computing. Robotic service hubs contain larger maintenance and logistics capabilities. Human-rated hubs exist at strategically important locations where people genuinely need to live, work, transfer between spacecraft or supervise complex operations.

The network does not need humans living at every hop.

People Could Live on the Moon and Mars

The Moon and Mars present a different case because permanent or semi-permanent settlements could eventually develop around scientific, industrial and infrastructure activities. Those people would not simply be "data-centre technicians." A lunar settlement might contain engineers, scientists, medical personnel, construction specialists, robotics experts, power engineers and resource-processing teams. Mars would require an even broader range of skills because communication delay makes the settlement more autonomous. AI and robots perform much of the physical work, but people become responsible for higher-level engineering, scientific judgement, governance, innovation and situations the autonomous systems were not designed to handle. This is another reason the argument that AI simply eliminates human work is incomplete. A machine civilisation that expands human capabilities can create entirely new environments in which human expertise becomes valuable.

Robots Could Become the Emergency Services of Space

Future infrastructure may therefore develop something analogous to emergency services. A platform stops responding. Nearby systems classify the incident. Traffic automatically fails over. A reconnaissance robot investigates. A repair vehicle launches. AI determines whether the platform can be recovered. Replacement equipment is dispatched. Humans intervene remotely if the autonomous system encounters something outside its approved procedures. In a human-occupied facility, emergency robots could also enter areas unsafe for people—fire, radiation exposure, depressurised compartments or damaged structures. The robots become not merely workers but first responders. This is particularly valuable in space because rescue can take far longer than on Earth.

Robots Must Eventually Build Larger Robots

The next stage is more ambitious. Suppose the lunar economy needs an excavator ten times larger than the machines originally delivered from Earth. Launching the completed excavator may be inefficient. Instead, lunar factories produce structural sections. Earth supplies precision components. Existing robots assemble them. The new machine can then extract resources at a larger scale. Those resources support larger factories. Those factories manufacture more components.

This creates a positive industrial feedback loop:

Robots build infrastructure → infrastructure produces materials → materials build larger machines → larger machines expand infrastructure. This is what I mean by robots building the machines that build the future. It is not necessarily a science-fiction scenario in which robots independently reproduce themselves without human involvement. It is a controlled industrial process in which automation allows manufacturing capacity to scale far beyond what astronauts could physically construct themselves.

The Difference Between Self-Repair and Self-Replication

This distinction is important because discussions of advanced robots can quickly become confused with fears of uncontrolled machine reproduction. Self-repair means a system can diagnose and replace some failed components. Self-assembly means machines can assemble equipment from supplied components. Local manufacturing means some components can be produced from available raw materials. Self-replication would mean a machine or industrial system could independently reproduce essentially all the components necessary to create another complete version of itself. These are very different levels of capability. The architecture in this book does not require uncontrolled self-replicating robots. For a long time, critical processors, specialised sensors and other components could remain dependent on controlled manufacturing systems and human-authorised supply chains. That provides both an engineering constraint and a governance mechanism.

AI and Robots Need Each Other

AI without robotics can analyse the physical world but has limited ability to change it. Robotics without sufficient intelligence requires humans to specify enormous numbers of actions. Combining them creates something much more useful. AI identifies a failing solar panel. A robot travels to it. Computer vision identifies the damaged module. The maintenance system confirms the replacement procedure. The robot removes the damaged component. A spare is installed. Electrical tests run. The AI confirms normal output. Earth receives the maintenance record later. The human operator did not need to control every movement, but humans designed the system, established its permissions and can review its actions. That is a much more realistic model of human-machine cooperation than either extreme of humans manually controlling everything or robots operating without governance.

The Ultimate Goal Is Infrastructure That Can Heal

A future Interplanetary Cloud cannot depend on every machine working perfectly forever. It must assume failure. Servers will fail. Robots will fail. Communication terminals will fail. Solar arrays will fail. Entire hops may occasionally fail. The architecture therefore needs to become increasingly self-healing. In computing, a self-healing system detects a failed service and automatically restores or replaces it.

The physical equivalent would operate at several levels:

Software fails → restart or move workload.

Server fails → switch to another server.

Module fails → robot replaces module.

Robot fails → another robot repairs or replaces it.

Hop fails → another hop takes over.

Orbital zone fails → traffic moves through another zone.

Only after continuity is restored does the repair process begin.

This is perhaps the most important principle:

Failover first. Repair second. Nobody running a critical terrestrial cloud service wants millions of customers waiting while an engineer repairs the failed server. Traffic moves elsewhere first. A civilisation depending on interplanetary infrastructure should operate the same way.

From Cloud Reliability to Civilisation Reliability

The concepts may sound futuristic, but their logic comes directly from systems engineering we already practise.

Redundancy. Availability zones. Health checks. Failover. Automated recovery. Predictive maintenance. Replaceable components. Disaster recovery. Distributed backups.

The new step is applying those principles not merely to software but to physical civilisation infrastructure. A future orbital node becomes something like an availability zone. A planetary compute region becomes something like a cloud region. Robotic maintenance becomes physical auto-healing. Interplanetary logistics becomes the supply chain behind the cloud. The analogy is not perfect, but it gives us a useful design philosophy: Do not try to build machines that never fail. Build a civilisation whose machines can fail without bringing civilisation down. That philosophy will matter enormously if people eventually live on the Moon, Mars and major orbital hubs. Their communications, AI, power systems, environmental controls and data infrastructure will not merely support websites or business applications.

Some will support human survival. The Interplanetary Cloud therefore cannot simply become bigger than today's cloud. It must eventually become far more resilient.

Chapter 23

Humans in the Interplanetary Cloud

The previous chapter described a future in which robots could eventually build, inspect and repair much of the physical infrastructure required beyond Earth. That raises an obvious question: if robots can perform so much of the physical work, why would humans need to be there at all?

The important word is **eventually**.

We do not currently possess autonomous machines capable of landing on Mars and independently constructing an entire industrial civilisation. Developing such capability would itself require a long progression involving smaller robotic systems, human engineering, orbital and lunar experience, increasingly capable autonomous machines and eventually industrial systems able to manufacture and maintain more of their own infrastructure.

Humans may therefore remain an important part of extraterrestrial development for a long time. AI and robotics can perform work that is dangerous, repetitive or difficult for people, while humans provide supervision, engineering judgement, scientific interpretation, complex repair, decision-making and intervention when machines encounter situations they were not designed to handle.

The First Humans May Arrive Before the First Settlement

There is an important distinction between **exploring Mars** and **permanently settling Mars**.

The robotic infrastructure described in this book—machines excavating terrain, extracting resources, constructing habitats, deploying large power systems and maintaining industrial facilities—belongs to a much later technological stage. Humanity does not yet possess fleets of autonomous machines capable of landing on Mars and independently building such infrastructure. Developing that capability could take decades.

That does not mean humans must wait until those machines exist before travelling to Mars.

Early human missions could have a very different purpose. They could conduct science, test life-support systems, study how humans cope with long-duration interplanetary travel and Martian gravity, operate experimental equipment, investigate local resources and learn how people and increasingly autonomous machines work together when immediate assistance from Earth is impossible.

Robotic missions already perform remarkable science on Mars. Rovers can examine rocks, analyse soil and minerals, study the atmosphere, search for evidence of ancient environments and collect samples. Future robotic systems will become considerably more capable.

The argument for sending humans therefore should not simply be that robots cannot examine Mars.

Humans still have an important advantage: they are highly adaptable general-purpose explorers. A scientist can notice something unexpected, change the day's objectives, investigate a nearby formation, improvise around an equipment failure or combine several observations in ways that were not anticipated years before the mission launched.

Whether that additional capability justifies the enormous cost and risk of a particular human mission is still a legitimate question. Sending people merely to plant a flag or repeat work that machines can already perform would be a weak foundation for a long-term Mars programme.

The greater value of early human expeditions may be the knowledge and operational experience they produce for what comes later.

They could teach us how habitats actually perform after years on Mars, how equipment responds to dust and temperature cycles, how reliably life-support systems can be repaired, how humans cooperate with autonomous machines and what kinds of failures were never predicted on Earth.

Those lessons could help engineers design the much more capable robotic systems required for the next stage.

Exploration Comes Before Industrial Preparation

The sequence proposed in this book is therefore not:

robots build an entire colony, and only then do humans visit Mars for the first time.

Nor is it:

humans arrive on Mars and manually construct an entire civilisation.

A more realistic progression could be:

Robots explore → small human expeditions explore and test systems → humans and robots learn together → autonomous industrial capability grows → machines prepare permanent infrastructure → sustained human settlement follows.

This distinction matters because the first human landing and the first permanent settlement may be separated by many years or even decades.

Early crews might live largely from equipment and supplies transported from Earth. Their bases could resemble temporary scientific and engineering outposts rather than self-sufficient cities.

Meanwhile, robotics would continue improving.

Eventually, machines could move from scientific exploration into large-scale physical work.

Machines Could Prepare the Permanent Settlement

Once autonomous robotics becomes sufficiently capable, the balance could change dramatically.

Robotic precursor missions would no longer merely photograph terrain or analyse rocks. Fleets of specialised machines could survey candidate settlement locations, map hazards, identify accessible resources and prepare landing areas.

Later generations could deploy communications infrastructure and power systems, move regolith, excavate sites, transport materials and assemble modular equipment delivered from Earth.

Mining systems could begin extracting useful local resources. Water-processing equipment could accumulate reserves where accessible water is available. Other systems could test the production and storage of oxygen and useful industrial materials.

Robots could assemble habitat structures and cover portions of them with local material for additional shielding. Maintenance machines could inspect equipment and replace standardised components.

Building such capability would itself be an enormous technological programme. It would require far more advanced robotics than today's planetary rovers, as well as reliable autonomous construction, energy systems, communications, resource processing and machine maintenance.

The objective would not necessarily be to build a complete city before another human ever travelled to Mars.

The objective would be different:

before large numbers of people attempt permanent settlement, machines should prepare as much of the life-supporting and industrial environment as reasonably possible.

The Moon Could Be the Proving Ground

Mars should probably not be the first place where humanity discovers whether large autonomous robotic industries actually work beyond Earth.

Earth will remain the starting point. Robots can be designed, trained and tested here under controlled conditions. Mining machines can be tested in deserts. Autonomous vehicles can operate across difficult terrain. Construction robots can practise assembling structures, while AI systems learn to coordinate fleets of machines.

But Earth can only test these technologies under Earth conditions.

Gravity, atmosphere, temperature, radiation, dust, communications, available resources and the difficulty of maintenance are different beyond Earth. A robotic system that performs perfectly in a terrestrial test facility has not yet demonstrated that it can become part of an extraterrestrial industrial system.

The Moon could therefore become an important proving ground.

It is close enough to Earth for communication to take only a little over a second each way, yet it presents a genuinely extraterrestrial environment. Machines would have to deal with vacuum, radiation, extreme temperature changes, abrasive lunar dust and reduced gravity. Equipment would also have to operate without technicians being able to walk into the facility whenever something fails.

That makes the Moon something more important than simply another destination.

It could become the **testing ground for the technologies eventually intended for Mars**.

Early lunar robots might perform relatively simple work: surveying terrain, moving material, inspecting equipment and deploying infrastructure. Later generations could attempt excavation, construction, resource processing, power-system maintenance and coordinated operation between multiple autonomous machines.

Failures would be valuable.

If an excavator repeatedly damages its joints because of lunar dust, engineers learn from it. If an autonomous construction system cannot align large modules reliably, the design changes. If robots cannot repair one another effectively, maintenance architecture improves. If communications fail, greater local autonomy becomes necessary.

Each generation could inform the next.

The progression might become:

Earth laboratory → terrestrial field testing → lunar deployment → lunar industrial experience → improved autonomous systems → Mars deployment.

The Moon also has an important advantage over Mars during this learning period: it is much closer.

Engineers on Earth could receive information quickly. Human crews could eventually intervene where necessary. Replacement equipment and upgraded machines would still be difficult and expensive to deliver, but the logistics problem would be far less severe than sending replacements to Mars.

This means lunar development could allow humanity to discover weaknesses in extraterrestrial robotics before those same weaknesses appear millions of kilometres farther away.

Settle the Moon Before Trying to Industrialize Mars

In the longer-term architecture proposed by this book, this suggests another possible sequence.

Humanity may first establish increasingly permanent infrastructure on and around the Moon. Humans and robots could initially work closely together. Lunar settlements could provide engineering workshops, communications facilities, power systems, laboratories and operational experience.

Robots would gradually perform more of the excavation, construction, inspection and maintenance.

Over decades, the Moon could become a place where humanity learns not merely how to **visit another world**, but how to **operate an industrial ecosystem on one**.

That distinction is critical.

A rover successfully travelling across Mars proves that we can operate an extraordinary machine on another planet. It does not yet prove that hundreds or thousands of machines can excavate resources, build infrastructure, distribute energy, repair one another and maintain an industrial system for decades.

Those are different levels of technological maturity.

Before relying on autonomous machines to prepare permanent Martian settlements, it would therefore make sense to demonstrate many of those capabilities closer to home.

Mars would not need to become the laboratory in which every major problem is discovered for the first time.

The Moon could be that laboratory.

And once robotic systems have demonstrated that they can build and maintain substantial infrastructure there, sending descendants of those machines to Mars becomes a much smaller conceptual leap.

The Moon would therefore serve two purposes in the development of the Interplanetary Cloud.

It could become humanity's first major extraterrestrial settlement.

But just as importantly, it could become the place where humanity learns how to build the machines that eventually make settlement of more distant worlds possible.

Test the Infrastructure Before Betting Human Lives on It

This is also a reliability problem.

Engineers do not normally place a critical production system into service simply because its individual components worked once during testing. They operate it, observe failures, measure performance and improve weaknesses.

A future Martian settlement should arguably be treated with even greater caution.

Before a long-duration human crew depended upon the infrastructure, robotic systems could demonstrate that the power network continues operating through difficult conditions. Communications could be tested over long periods. Water extraction and storage systems could demonstrate their actual production rates. Habitats could be checked for leaks and structural problems. Robots could practise replacing components. Spare parts could be positioned before they were urgently needed.

Failures during this period would still be expensive, but they would primarily endanger machines rather than people.

This creates an important principle for extraterrestrial settlement:

Build first. Test second. Send people when the system has demonstrated that it can support them.

The principle will never remove all risk. Human exploration has always involved uncertainty, and Mars will contain problems that cannot be completely predicted from Earth. But there is an enormous difference between accepting unavoidable exploration risk and deliberately exposing people to risks that machines could have reduced beforehand.

From an Empty Planet to an Operating Outpost

This changes what the first long-duration human arrival could look like.

Instead of astronauts landing beside cargo containers and immediately beginning the struggle to create basic infrastructure, they could arrive at an operating robotic outpost.

Power systems could already be generating electricity. Communications could already be functioning. Supplies could already be stored. Robotic equipment could already be moving material and maintaining infrastructure. Water and oxygen reserves could potentially have been accumulated where the technology and local resources allowed it. Habitats could have been deployed and monitored before people entered them.

The first crew would still face enormous risks and would still have substantial work to perform. But their role would increasingly become one of **commissioning, engineering, scientific work, expansion and solving exceptional problems**, rather than performing every basic physical task themselves.

In this model, the beginning of a Martian settlement may not be the moment the first human steps onto Mars.

It may begin years earlier, when the first network of machines starts preparing the place where humans will eventually live.

Why Humans Would Still Matter

If robots prepare so much of the infrastructure, another question follows: why send humans at all?

Because machines and humans have different strengths.

AI can process enormous quantities of information. Robots can repeat physical operations accurately and work in places that are dangerous to human biology. But a complex extraterrestrial settlement will inevitably encounter situations its designers did not predict.

A pipe may fail in an unexpected way. Dust may enter equipment despite protective systems. Two individually minor faults may interact to create a problem nobody modelled. A robot may be capable of replacing a standard module but unable to remove a component that has become physically deformed. Scientific discoveries may change priorities immediately.

Humans are unusually capable general-purpose problem solvers. We can combine physical manipulation, judgement, improvisation, experience and understanding of broader objectives.

This suggests that early extraterrestrial settlements may contain relatively small numbers of highly capable people supported by much larger machine ecosystems.

The human population would not need to scale in direct proportion to the amount of infrastructure.

Ten times more mining equipment should not necessarily require ten times more miners. A larger solar installation should not automatically require a proportionally larger maintenance workforce. AI, autonomous machines and remote engineering could allow a relatively small human population to supervise an industrial system far larger than anything those people could operate manually.

A Different Kind of Space Worker

This also changes what it means to work in space.

A future Mars infrastructure engineer may never live permanently on Mars.

One engineer could work from Earth, another from an orbital facility and another during a temporary rotation on the Martian surface. AI systems could monitor thousands of machines continuously. Local robots could perform routine inspections and repairs. Humans would become involved when judgement, authorisation or unusual engineering intervention was required.

Other specialists might rotate between locations rather than permanently migrate. Engineers, doctors, scientists, construction specialists and equipment technicians could spend months at an extraterrestrial facility and later return to Earth.

Eventually, some people would probably choose permanent settlement. Families, communities, institutions and local cultures could emerge. At that point the purpose would no longer be simply operating machines. A genuine extension of human civilisation would have begun.

But permanent human settlement does not require humans to perform every physical task themselves.

Human Settlements and Machine Civilization Can Grow Together

The result may therefore be neither a Solar System populated everywhere by humans, nor one surrendered entirely to autonomous machines.

Different locations would develop differently.

Earth would remain the biological and economic centre for a long time. The Moon could contain increasingly sophisticated human-machine industrial facilities. Mars could develop robotic infrastructure first and human settlements later. Large orbital facilities might support rotating or permanent crews where their presence provides genuine value. More hostile destinations could remain almost entirely robotic.

This is not human retreat.

It is division of capability.

Humans require relatively narrow environmental conditions but bring judgement, values, creativity, curiosity and social organisation. Machines can operate in environments hostile to biology but require energy, maintenance, objectives and governance.

There is therefore no reason to force both into identical roles.

Let robots excavate dangerous terrain. Let machines inspect exposed infrastructure. Let AI optimise enormous networks. Let humans decide why those networks exist. Let robots handle predictable failures. Let humans investigate the unpredictable.

The more useful model is not humans **versus** machines.

It is humans and machines building an ecosystem in which each operates where its capabilities make the most sense.

From Fear of Replacement to Expansion of Possibility

At the beginning of this book, AI appeared primarily as a disruptive force. It could write code, automate support work, monitor infrastructure and perform tasks that once required larger human teams.

That disruption is real.

But the story does not have to end there.

The same technologies reducing labour requirements in one part of today's economy could make entirely new industries possible elsewhere. AI can operate systems too complex for continuous manual control. Robots can work where humans cannot safely work. Automation can maintain infrastructure too distant for conventional support.

And before humans settle another world, those technologies could prepare it.

That creates work on Earth as well as beyond it. Engineers would design autonomous construction systems. Mining specialists would develop extraterrestrial extraction methods. Platform and reliability engineers would operate distributed machine fleets. Communications specialists would build planetary networks. Manufacturing industries would produce modular equipment. Scientists would determine where infrastructure should be placed. Human crews would eventually commission, repair and expand what those systems had constructed.

The question should therefore not only be:

Which existing jobs will AI replace?

It should also be:

What can humanity build with AI that was previously impossible—and what new human work appears when we decide to build it?

The Interplanetary Cloud is one possible answer.

Its purpose is not to send humans away because machines have taken their jobs. Its purpose is almost the reverse. AI and robotics could expand the physical frontier within which human civilisation can create meaningful work, knowledge and opportunity.

The first machines on Mars may therefore matter for something much larger than collecting scientific data.

They may be the first construction workforce of a future human world.

And once humans and machines begin operating together, another boundary appears. There will still be destinations where maintaining human life makes little practical sense at all. Those places may belong primarily to autonomous infrastructure.

That is the next stage of the architecture.

Chapter 24

Where Humans Cannot Go: Autonomous Infrastructure Beyond Earth

The previous chapter considered how humans might work alongside AI and robots on the Moon, Mars and selected orbital facilities. But human presence should not become the default objective for every location in the Solar System.

Some environments may be so hot, cold, irradiated, remote or otherwise hostile that keeping people alive there would require far more infrastructure than the work itself.

This is where AI and robotics fundamentally change the architecture.

Civilisation does not need to make every destination suitable for humans before it can make productive use of it.

Some locations may contain humans and machines working together. Others may be visited occasionally by people. Some could remain almost entirely robotic.

The important question therefore changes from:

Can humans live there?

to:

Can we build an autonomous system capable of operating there safely?

Machines Can Go Where Biology Struggles

Humans operate within relatively narrow biological limits. We require breathable air, suitable pressure, water, food, manageable temperatures and protection from radiation. Long periods in low gravity or microgravity also affect the human body.

Machines have different limitations. Robots still require energy, thermal control, suitable materials, radiation-tolerant electronics, communications and maintenance. They are certainly not indestructible.

But keeping a machine operating in a hostile environment can sometimes be much easier than keeping both the machine **and an artificial Earth-like environment around its human operator**.

Mercury provides a useful example. Its surface experiences extreme temperature differences and an intense solar environment. Establishing a conventional human industrial settlement there would require extraordinary protection.

A future specialised robotic facility would still face serious engineering problems, but it would not require breathable air, food production or a comfortable living environment. Robots could inspect equipment, local computing could make immediate operational decisions, and humans elsewhere could provide higher-level instructions.

Venus makes the distinction even clearer. Its surface temperature and atmospheric pressure make conventional human habitation extraordinarily difficult. Machines designed for such conditions would also face severe material, electronic and thermal challenges, but robotics at least allows us to consider operating there without first constructing an entire protected human settlement.

The principle is simple:

A machine can become our physical presence somewhere our bodies cannot reasonably operate.

The machine does not need to look human either. A lunar mining robot might resemble an excavator. An orbital repair system could consist of articulated arms around a central platform. An asteroid-mining machine might attach itself directly to the rock. A Venus atmospheric system might be an aircraft or balloon.

The form should follow the environment.

AI provides intelligence. The robot provides the physical interface through which that intelligence interacts with the environment.

Human Settlement Is Not the Measure of Expansion

This changes what becoming an interplanetary civilisation could mean.

Science fiction often measures expansion by the number of places where humans establish cities. But the physical reach of civilisation could become much larger than the area permanently inhabited by people.

Earth could remain humanity's primary biological home. The Moon could contain human facilities alongside much larger robotic industrial regions. Mars could combine rotating or longer-duration human crews with increasingly autonomous infrastructure. Orbital hubs could vary from heavily automated facilities to major stations supporting people.

Asteroids and some extreme planetary environments might contain almost no permanent human population at all.

This is not a limitation of expansion.

It is another form of expansion.

AI and robotics become our reach.

The Solar System could therefore develop as a mixture of biological and machine environments rather than requiring humans to reproduce Earth-like settlements everywhere civilisation operates.

Distance Changes the Meaning of Control

Once machines operate far from people, however, another problem appears.

Who controls them?

On Earth, a remotely operated machine can often receive instructions almost immediately. The Moon introduces a noticeable but manageable communication delay. Mars requires delays measured in minutes, depending on the positions of the planets. More distant destinations create still longer delays.

A human operator therefore cannot continuously control every movement of a machine operating far across the Solar System.

The relationship gradually changes:

Remote control → supervised autonomy → local autonomy

With remote control, a human directs much of the machine's activity.

With supervised autonomy, humans define a task while the machine determines many of the individual actions required to complete it.

With greater local autonomy, humans increasingly define objectives, permissions and safety boundaries while local AI handles immediate execution.

This does not mean humans disappear from the system.

It means human control moves to a different level.

Instead of saying:

Move this robotic arm ten centimetres to the left.

a future engineer might say:

Inspect Solar Array 14, identify damaged modules, repair faults that fall within your approved procedures and report anything outside your authority.

That distinction becomes essential as distance increases.

Autonomy Does Not Mean Unlimited Authority

Giving machines greater independence creates an obvious risk. A robot capable of acting without immediate human instructions must not interpret autonomy as permission to do anything it considers useful.

Autonomous infrastructure therefore requires **governance built into the system itself**.

Imagine a remote robotic mining facility. Its machines may be authorised to inspect equipment, replace standard components and adjust routine operations independently. But they might not be authorised to open an entirely new mining area, substantially alter the environment or change the purpose of the facility without human approval.

An orbital station might be authorised to perform an emergency collision-avoidance manoeuvre immediately because waiting for permission could be dangerous. The same system might require human approval before making a major planned change to its orbit or mission.

Different actions therefore receive different levels of authority.

This produces an important principle:

Autonomy does not eliminate governance. It makes predefined governance executable when humans are unavailable.

Humans establish the boundaries. Machines operate within them.

The Machine Needs a Safe Mode

What happens when something goes wrong and no human can respond immediately?

Spacecraft already use versions of a concept known as **safe mode**. When an unexpected condition occurs, the system enters a protected operating state intended to preserve essential functions while reducing unnecessary activity.

Future autonomous infrastructure would require much more sophisticated versions of this principle.

Imagine a robotic mine that suddenly loses communication with its control centre.

The correct response may not be to continue operating normally. But shutting down absolutely everything could also create danger. Thermal management may need to continue. Energy systems may need to remain stable. Mobile equipment may need to move into safe positions.

The facility could therefore enter a predefined state.

Mining stops. Dangerous equipment is secured. Essential power and thermal systems continue. The communication network attempts alternative routes. Local AI investigates the failure and records what happened.

When communication returns, human engineers receive the diagnostic information and determine whether normal operations should resume.

The machine does not blindly continue.

But neither does every communication failure cause the entire facility to collapse.

It moves into a known, controlled state.

Local AI Becomes the First Line of Support

This also changes how technical support works.

Today, when something fails inside a terrestrial data centre or industrial facility, people can often reach the equipment relatively quickly. At a distant extraterrestrial facility, physical assistance may be hours, days or months away.

Support therefore moves progressively closer to the machine itself.

A future system could use an escalation model such as:

Level 0 — Automatic software recovery

The system detects a familiar problem and corrects it automatically.

Level 1 — Local AI diagnosis

AI investigates the fault, checks telemetry and determines whether an approved recovery procedure exists.

Level 2 — Robotic inspection and repair

A maintenance robot physically examines the equipment and performs repairs within its authorised capability.

Level 3 — Remote human engineering support

If local systems cannot resolve the problem, engineers elsewhere analyse the information and provide higher-level instructions.

Level 4 — Physical replacement or recovery mission

If the problem cannot be solved remotely or robotically, replacement equipment or human intervention may eventually be required.

Distance determines how heavily infrastructure depends on the earlier levels.

On Earth, physical intervention may mean someone walks into a data centre. On the Moon, it might involve a nearby robot or human crew. On Mars, assistance from another location could take considerably longer. At a distant asteroid, outside physical help might be months away.

The farther infrastructure moves from immediate human support, the more capable its local diagnosis, recovery and maintenance systems must become.

AI Must Know When It Does Not Know

Increasing autonomy creates another requirement that may be even more important than the ability to act:

the ability not to act.

An autonomous system should have an approved **operating envelope**—the conditions and situations within which it has been designed, tested and authorised to make decisions.

Imagine a maintenance AI trained to diagnose thousands of known equipment failures. It encounters a fault it has never seen before.

An intelligent system should not assume that because it can generate a possible solution, it has permission to try it.

The safer response may be:

I do not have sufficient confidence to perform this repair. I will isolate the affected equipment, transfer its workload where possible and request assistance.

That ability becomes increasingly important as AI controls physical systems.

A wrong answer from a chatbot may produce incorrect information. A wrong decision by a machine controlling a mining vehicle, spacecraft, power system or industrial robot can damage physical infrastructure or endanger people.

Trustworthy autonomy therefore requires more than intelligence.

It requires boundaries, confidence thresholds, fallback procedures and escalation.

Sometimes intelligence means knowing what to do.

Sometimes intelligence means recognising when not to act.

Autonomous Does Not Mean Human-Free

There is an important distinction between infrastructure that can **operate autonomously** and infrastructure that never requires humans.

A Mars facility may operate for long periods using local AI and robots while still depending upon human engineers for unusual failures, major upgrades, governance and strategic decisions.

A lunar industrial region might contain hundreds or thousands of autonomous machines while a much smaller rotating human workforce supervises the overall system.

A remote asteroid operation might function without any people physically present, while engineers on Earth or elsewhere remain responsible for its objectives and behaviour.

The balance would depend on distance, danger, economics and technological maturity.

As discussed earlier, humans may be especially important during the establishment of early extraterrestrial industrial ecosystems. Robots can survey sites and prepare initial infrastructure, but people may still be required to solve unexpected problems, repair systems and help construct the industrial capability from which much greater machine autonomy later develops.

Over time, the balance can shift:

Human-intensive establishment → human-supervised automation → increasingly autonomous operation

The objective is therefore not to remove humans from civilisation.

It is to stop requiring human bodies to be physically present for every useful activity civilisation performs.

A Distributed Civilisation Needs Distributed Responsibility

Once autonomous machines operate across many locations, responsibility must remain clear.

AI may decide how to execute a task, but humans and institutions still need to determine why that task exists, what risks are acceptable, what resources may be used and what actions are prohibited.

This becomes especially important if machines eventually operate industrial facilities far from Earth.

A mining system should not simply expand indefinitely because additional material is available. A robotic construction network should not redesign its mission because doing so appears locally efficient. A powerful autonomous system should not acquire unlimited authority merely because communication with humans is inconvenient.

The farther machines move from us physically, the more carefully their permissions may need to be designed before deployment.

The architecture therefore becomes:

Human objectives → defined authority → local AI decisions → robotic action → monitoring → escalation when necessary

Distance changes where decisions are made.

It should not eliminate accountability for those decisions.

Humans and Machines Occupy Different Parts of the Map

This produces a different relationship between humanity and AI from the one often imagined in debates about automation.

Humans and machines do not necessarily have to compete for the same jobs, the same buildings or eventually even the same physical environments.

They can specialise.

Humans remain concentrated where biological civilisation can flourish. Machines increasingly operate where human biology becomes a disadvantage. AI coordinates those machines. Humans establish objectives, laws and acceptable risks. Robots extend our physical reach, and the knowledge they gather flows back through the wider network.

The story therefore changes from:

AI replaces humans

to:

AI allows humanity to operate where humans alone cannot.

This is one way the technological disruption discussed at the beginning of this book could eventually become technological expansion.

Earth Remains the Environment We Protect Most

There is also a larger consequence.

The farther we explore the Solar System, the more obvious Earth's biological value becomes.

Mercury does not provide breathable air. Venus does not. The Moon does not. Mars does not. Space itself certainly does not.

Earth already provides the extraordinarily complex environmental system that every extraterrestrial human habitat would have to imitate artificially.

Expanding beyond Earth should therefore never become an excuse to neglect it.

The opposite principle makes more sense.

Earth is the environment we protect precisely because machines can increasingly operate elsewhere.

Selected computing, robotic industry, resource extraction and experimental infrastructure could gradually gain access to environments where permanent human populations are unnecessary. Earth can remain the biological centre of civilisation rather than simply becoming the default location for every additional machine humanity builds.

But once machines begin operating throughout these environments, another fundamental problem appears.

Every one of them needs energy.

A mining robot needs power. A communication node needs power. A lunar data centre could eventually require enormous amounts of power. A Martian compute region may require far more.

Unlike Earth, these facilities cannot simply connect to one national electricity grid.

Fortunately, almost every world and spacecraft in this architecture has access, directly or indirectly, to the same extraordinary energy source:

the Sun.

The difficulty is that planets rotate, spacecraft orbit, shadows move, Mars receives less sunlight than Earth, and some facilities periodically lose direct sunlight altogether.

If computing eventually spreads through the Solar System, infrastructure may need to learn to follow available energy rather than assuming energy will always follow the computing.

That brings us to the next stage:

Follow-the-Sun Computing, Literally.

Chapter 25

Follow-the-Sun Computing, Literally

Every robot, communication relay, habitat and data centre in the Interplanetary Cloud needs energy.

On Earth, we normally decide where computing should operate and then build an electricity system capable of supplying it. Beyond Earth, that relationship may sometimes reverse.

Computing may need to adapt to where energy is available.

The Sun is the largest readily accessible energy source across much of the Solar System. Solar panels already power satellites, spacecraft, the International Space Station and Mars rovers. But a future Interplanetary Cloud should not depend on solar energy alone. Different locations may require combinations of solar generation, nuclear power, energy storage and technologies that may emerge much later.

The important principle is therefore not:

Everything must run on solar power.

It is:

Computing should understand the physical environment in which it operates.

Different Worlds, Different Energy

Solar energy is not equally available throughout the Solar System.

As distance from the Sun increases, the same energy spreads across a larger area. Mars therefore receives less solar energy than Earth, while locations closer to the Sun receive considerably more.

But distance is only part of the problem. The Moon experiences long periods of daylight and darkness across much of its surface. Mars can experience dust conditions that reduce solar generation. Some orbital locations may receive sunlight for long periods but still require energy storage for interruptions.

There is therefore no single extraterrestrial power architecture.

A lunar data centre might combine solar generation, storage and nuclear power. Martian infrastructure may require an even more diverse energy mix. Orbital computing facilities could be designed around the particular sunlight conditions of their orbits.

The architecture must follow the environment.

Not Every Computation Has to Run Now

The other side of the equation is the workload itself.

Some computing cannot wait.

Life support must continue operating. A spacecraft cannot postpone collision avoidance. Robots performing dangerous operations need immediate local control. Navigation and emergency communication must remain available.

Other workloads are much more flexible.

AI training, scientific simulations, archival processing, large analytical jobs and some forms of data replication can often be delayed without serious consequences.

These are **deferrable workloads**—computational jobs that can move in time, and sometimes location, without making the system unusable.

Imagine a lunar data centre with abundant solar power during part of its operating cycle. It could use that period for energy-intensive AI training or scientific processing. When available generation falls, those workloads could slow or pause while critical systems continue.

The data centre would therefore treat computing demand as something that can be managed rather than assuming every processor must run at maximum capacity continuously.

Compute Can Follow Energy

Now extend that principle across the Interplanetary Cloud.

Future computing facilities around Earth, the Moon and Mars would not experience identical energy conditions. Some locations may temporarily have abundant generation and spare computing capacity while others are operating under tighter constraints.

A future scheduler could therefore ask:

Where can this workload be performed most efficiently right now?

If a computational task does not have to run at a particular location, software could consider energy availability, computing capacity, communication requirements and the location of the required data before deciding where to execute it.

This is **energy-aware workload scheduling**.

Instead of always building additional electricity generation wherever computing happens to be located, flexible computation can sometimes move toward available energy.

In simple terms:

Sometimes it may be easier to move the computation than to move the electricity.

Distance Still Wins When Latency Matters

This flexibility has strict limits.

Earth and Mars are separated by enormous distances. Moving a workload to another planet makes no sense if the application requires an immediate response.

A Mars habitat's oxygen-control system must remain on or near Mars. A spacecraft's navigation system must remain with the spacecraft. Autonomous robots require enough local intelligence to operate safely without waiting for instructions from Earth.

Large datasets create another limitation.

Engineers sometimes use the term **data gravity** to describe the tendency for computation to remain close to very large collections of data because moving the entire dataset may cost more time and resources than moving the computation.

If Mars eventually contains enormous collections of geological and scientific information, transmitting all of that data toward Earth merely to analyse it may make little sense. The analysis should often run on Mars instead.

The scheduler therefore has to balance several things:

Energy + Compute Capacity + Data Location + Communication Capacity + Latency + Urgency

This produces a simple rule:

Latency-sensitive computing stays local. Flexible computing can move.

Energy Diversity Still Matters

Moving flexible workloads does not eliminate the need for reliable local power.

Critical infrastructure cannot depend on the assumption that another planet will always perform its computing. Each important location therefore needs its own resilient energy architecture.

Solar generation may provide one layer. Energy storage can bridge interruptions. Nuclear systems could provide stable generation where sunlight is intermittent or insufficient. Future technologies may eventually add additional options.

Orbital solar generation and even directed power transmission using microwaves or lasers could someday provide specialised supplementary energy, although such systems would introduce conversion losses, infrastructure requirements and significant engineering and safety challenges.

The principle is therefore:

Generate locally where practical, store energy where useful, and move flexible computing when doing so provides an advantage.

Computing Must Follow Heat Too

Energy availability alone does not determine the best location for computing.

Every watt consumed by computing eventually becomes heat.

Space may appear cold, but a data centre in a vacuum cannot rely on surrounding air to remove heat. Much of its waste heat ultimately has to be rejected through thermal radiation using radiator surfaces.

More computing therefore means more electricity and more heat, which requires greater thermal-management capacity.

A location with abundant energy is not automatically an ideal location for a data centre.

The scheduler of the future may consequently need to understand not only available electricity but also whether a facility has sufficient cooling capacity to accept additional work.

Follow the Resources

This reveals that **Follow-the-Sun Computing** is really part of a larger concept.

The future system does not merely ask:

Where is sunlight available?

It asks:

Where are the resources needed to perform this computation most effectively?

Sometimes the answer depends on energy.

Sometimes it depends on cooling.

Sometimes it depends on where the data is stored.

Sometimes it depends on available AI accelerators.

Sometimes communication delay makes moving the workload impossible.

The deeper principle is therefore:

Move flexible digital work toward the resources that can perform it most effectively and sustainably.

This is **Follow-the-Resources Computing**.

We Can Begin on Earth

None of this requires waiting for lunar or Martian data centres.

The same principle can already be applied on Earth. Where operationally appropriate, delay-tolerant computing can be scheduled around periods or locations with greater availability of lower-carbon electricity.

AI training does not always need to begin immediately. Batch processing can sometimes wait. Certain workloads may be able to move between computing regions, provided security, data residency, network capacity and latency requirements permit it.

Before consuming more energy, the system can ask a surprisingly powerful question:

Does this computation really have to run here, right now?

As computing expands beyond Earth, that question becomes even more important.

The Interplanetary Cloud should not simply be a collection of data centres scattered through space. It should become an infrastructure capable of understanding the physical resources beneath the digital world.

Critical computing stays where it must. Flexible computing follows the resources.

Chapter 26

Data Has a Place: Earth, the Moon and Mars as Storage Tiers

Modern computing depends on redundancy. Important information is rarely stored in only one place. A bank, government, hospital or technology company may keep several copies of its data across different servers, data centres and geographic regions. If one system fails, another copy remains available.

This architecture is essential, but it creates a growing physical problem. As humanity produces more data—and as AI systems generate, analyse and retain enormous datasets—we do not only need capacity for the original information. We need capacity for its replicas, backups, historical versions and disaster-recovery copies.

That means data growth can multiply infrastructure growth.

The Interplanetary Cloud offers another possibility. Instead of asking whether data centres should simply move from Earth to space, we can ask a more useful question:

Which data actually needs to remain close to Earth?

Latency Determines Where Data Can Live

Not all information has the same urgency.

Consider a banking transaction. When someone transfers money, pays for something or withdraws cash, the system needs to know the current account state almost immediately. Waiting several minutes to contact another planet before approving the transaction would obviously be unacceptable.

Such systems belong on Earth or, in a more advanced future, in orbital computing infrastructure sufficiently close to Earth to provide very low communication latency.

Now consider a secondary copy of the same bank's historical records. Transactions from five or ten years ago remain important, but they normally do not need to be retrieved within milliseconds.

The physical location of that data can therefore be different.

This gives the Interplanetary Cloud a simple architectural principle:

The more urgently data must be accessed, the closer it should remain to the users who need it.

Earth and Earth Orbit: The Real-Time Layer

Earth would remain the primary computing region for services requiring immediate interaction with people and terrestrial machines.

Payment systems, financial trading, authentication, emergency services, industrial control, interactive applications and many other operational workloads would continue to depend on extremely low latency.

Future orbital data centres around Earth might supplement this infrastructure, but the fundamental requirement remains the same: real-time data stays close to where real-time decisions occur.

Space therefore does not replace terrestrial cloud computing.

It adds new tiers to it.

The Moon: The Near-Earth Resilience Layer

The Moon provides an interesting intermediate location.

Light takes only about 1.3 seconds to travel between Earth and the Moon. A request requiring communication in both directions would therefore experience at least a few seconds of physical delay once transmission and processing are considered.

That is far too slow for many real-time transactions, but it could be perfectly acceptable for numerous backup and disaster-recovery functions.

A future organisation might therefore operate its primary database on Earth while maintaining an asynchronous secondary replica on the Moon.

Asynchronous replication means the primary system does not wait for the distant copy to confirm every update before continuing. Changes are transmitted afterward.

A banking transaction could therefore complete on Earth immediately. The updated record could subsequently be replicated to lunar infrastructure.

The Moon would not run the transaction itself. It would provide another physically independent copy.

Mars: The Deep-Archive Layer

Mars belongs to an entirely different category.

Depending on the positions of Earth and Mars in their orbits, communication takes several to more than twenty minutes in one direction. A request-and-response retrieval could therefore take tens of minutes and, under some conditions and with processing overhead, approach an hour.

That makes Mars unsuitable as a conventional extension of an Earth application.

But historical information does not care very much about latency.

Scientific archives, historical records, completed research datasets, older financial records, cultural material, engineering documentation, previous AI model versions and many other datasets may remain valuable for decades or centuries while being accessed only occasionally.

For these workloads, waiting tens of minutes for retrieval may be completely acceptable.

Mars could therefore eventually function as a **deep-storage region** rather than a normal cloud region.

The architecture becomes:

Earth / Earth Orbit → Real-time operational data

Moon → Secondary replicas and near-Earth resilience

Mars → Historical archives and deep resilience

Distance stops being merely a disadvantage. It becomes part of the storage architecture.

Not Every Copy Needs an Earth Data Centre

This distinction becomes increasingly important as AI expands.

AI does not merely consume information. It can generate enormous quantities of logs, model outputs, synthetic data, intermediate results, training datasets and new digital content. Meanwhile, organisations continue producing conventional business, scientific and human-generated information.

Every important dataset may also produce several additional copies.

If all of those copies must remain inside terrestrial data centres, humanity may repeatedly build new facilities simply to store redundant information.

But redundancy does not necessarily require proximity.

A live database may need two or three extremely fast copies near its users. Its older snapshots may not. A scientific dataset actively being analysed may require high-performance storage on Earth, while completed observations could eventually move into slower archival tiers. An AI model being actively trained might require enormous terrestrial or orbital computing resources, while previous generations of that model could be preserved much farther away.

The objective is therefore not:

Move Earth's data centres to Mars.

It is:

Put each category of data at the closest location its latency requirements actually justify.

A Planetary Version of Storage Tiering

We already use this principle inside modern computing.

Frequently accessed information can be kept on extremely fast storage. Less frequently accessed information can move to cheaper and slower systems. Long-term archives can be placed on storage designed primarily for preservation rather than immediate performance.

The Interplanetary Cloud extends the same principle across physical distance.

Instead of only:

Fast storage → slower storage → archival storage

a future civilisation could eventually have:

Earth → Earth Orbit → Moon → Mars

Each step trades immediate accessibility for greater physical separation and potentially different resource and resilience characteristics.

This does not eliminate the need for redundancy on Earth. Critical systems would still require multiple terrestrial copies because waiting for the Moon or Mars whenever a server fails would make no sense.

But it could change how much long-term redundant capacity Earth ultimately needs to carry.

Distance Can Become a Feature

Today, distance is usually treated as the enemy of computing.

For real-time applications, it is.

For archival information, however, distance may become useful.

The Moon can provide a physically separate resilience layer while remaining only seconds away. Mars can provide much deeper separation for information that can tolerate retrieval delays measured in tens of minutes.

The important question is therefore no longer simply:

Where can we build another data centre?

It becomes:

How quickly does this particular information actually need to come back?

Once we ask that question, the Solar System begins to look less like a collection of impossibly distant locations and more like a hierarchy of future storage tiers.

Earth remains the real-time computing centre of human civilisation. Earth orbit can expand nearby computing capacity. The Moon can become a secondary resilience layer. Mars can become a deep historical archive.

As AI causes civilisation's information footprint to expand, this architecture could eventually reduce the pressure to keep building terrestrial data centres merely to hold additional redundant copies of information that rarely needs immediate access.

The speed of light remains an unavoidable limitation.

But not every byte needs to be fast.

Chapter 27

The Solar System Data Explosion

At the beginning of this book, the data problem was primarily terrestrial. Humans create documents, photographs, videos, transactions and messages. Companies generate telemetry. Sensors continuously measure machines, while AI systems consume enormous datasets and create additional information of their own. Now imagine extending that activity across the Solar System.

A future lunar mining operation could contain thousands of sensors. Robots would continuously monitor their motors, batteries, navigation systems and surroundings. Scientific instruments would observe the Moon. Habitats would monitor life support, agriculture and energy. Spacecraft would generate navigation and engineering information. Communication relays and data centres would produce their own operational records. Mars would eventually do the same. As thousands—and perhaps eventually millions—of autonomous machines become active, the Interplanetary Cloud would become an enormous distributed sensor network.

This creates a paradox: **the more successfully civilisation expands computing into the Solar System, the more data that expansion itself creates.** The answer cannot simply be to keep building more storage.

Consider a single lunar excavator. Its cameras may continuously record images while its motors report temperature and electrical current, its batteries report their condition, its navigation system records movement and its mining sensors analyse material. These continuous measurements are called **telemetry**. AI then creates another layer of information by analysing that telemetry. Instead of merely recording vibration from a pump, for example, an AI system might conclude that its bearing is deteriorating and recommend maintenance within the next few weeks. We may now possess the original sensor readings, processed information, AI analysis, maintenance decision and audit record—all originating from one event.

Multiply this by millions of machines and the scale becomes obvious.

It would make little sense to send all of this raw information back to Earth. Imagine thousands of cameras operating continuously in a Martian industrial region. Transmitting every frame across interplanetary space simply because Earth has storage capacity would consume enormous communication resources. Much of the processing therefore needs to happen where the information is created. This is **edge computing**—processing data close to its source rather than automatically sending everything to a distant data centre.

A robot inspecting ten kilometres of pipeline, for example, might record hours of video in which almost nothing happens. Local AI could analyse that video, identify three suspicious sections, retain the important high-resolution footage and produce a verified inspection report. Depending on engineering and legal requirements, the remaining routine footage could eventually be deleted.

The architecture becomes simple:

Sense → Process locally → Identify significance → Preserve what matters → Transmit what is useful

This creates a much more important question than how much storage civilisation can build: **what information actually deserves to be preserved?**

At Solar System scale, humans could not manually make that decision for every file produced by millions of machines. AI may therefore become a **data curator**, operating according to rules established by humans and organisations. Critical engineering records could be retained. Scientific observations could be archived. Routine telemetry could be summarised. Duplicate information could be removed. Temporary diagnostic records could expire automatically, while an accident or unusual scientific observation could immediately trigger preservation of all original evidence.

Every piece of information therefore needs a **data lifecycle**. Instead of assuming that creation automatically means permanent storage, information might move through a sequence such as:

Create → Process → Classify → Store → Archive or Delete

Different information would follow different paths. A scientific observation might remain valuable for centuries. Evidence from an equipment failure could be preserved for future engineering analysis. Routine temperature readings from a machine operating normally might eventually be reduced to statistical summaries. Temporary diagnostic information might become useless once a fault has been resolved.

AI could also reduce data intelligently, but this introduces an important danger. **A summary is not the same as the original evidence.** If twelve hours of inspection video are reduced to a short report, something that appears irrelevant today might turn out to be important years later. Critical scientific observations, safety incidents and unusual events may therefore require preservation of the original data, while routine operations may permit much more aggressive summarisation and deletion. The objective is not to delete as much as possible. It is to distinguish valuable information from digital noise.

The same principle applies to backups. Replication protects information from failure, but copying everything repeatedly also multiplies storage requirements. One exabyte copied ten times requires ten exabytes of storage before additional overhead is considered. Critical information may justify several independent copies, while temporary telemetry may require none after it has been processed. The goal is not maximum replication but **sufficient resilience according to the value and risk of the data**.

Chapter 26 showed how those necessary copies can occupy different locations according to latency and resilience requirements. Active information may need to remain near Earth or its users, while less time-sensitive copies can exist farther away. The important point here is different: **not every piece of information deserves the same number of copies, and some information does not need long-term preservation at all.**

This matters because digital storage is ultimately physical. Every stored byte eventually depends on hardware somewhere. Storage devices require materials and manufacturing. They consume electricity, produce heat, occupy infrastructure and eventually need replacement. Replication multiplies those requirements. There is therefore **no truly weightless cloud; there are only computers somewhere else.**

A civilisation producing data at enormous scale must consequently become good not only at remembering, but also at forgetting. Deletion becomes an engineering feature rather than a failure. Routine telemetry may be summarised and removed. Duplicate information may disappear once verified copies exist. Temporary data may expire. Critical knowledge, scientific evidence and historically important information may remain for generations.

Humans would still need to determine the rules. Questions involving medical records, scientific evidence, personal information, legal requirements and civilisation archives cannot simply be delegated to an AI system without governance. AI can apply retention policies across billions of files, but people and institutions must decide what those policies are intended to protect.

The Solar System data problem is therefore not ultimately:

How do we build enough storage to keep everything?

It is:

How do we decide what deserves to survive?

The Interplanetary Cloud will need to become selective from the moment information is created. Process locally. Transmit what is useful. Preserve original evidence when its future value is uncertain. Replicate according to risk. Delete information when keeping it no longer justifies the physical resources required to preserve it.

Because a civilisation capable of generating almost unlimited data must eventually learn something equally important:

what not to preserve.

Chapter 28

Data Centres Become Space Stations

A future orbital data centre may begin with a narrow purpose: provide computing, storage and communications outside Earth. But once humanity has paid the enormous cost of placing a large, powered structure in space, maintaining its orbit, controlling its temperature and protecting its electronics, it makes little sense for that structure to perform only one function. The same infrastructure could gradually support communications, spacecraft docking, robotic maintenance, cargo handling and eventually temporary human accommodation. In this way, some data centres could evolve into something closer to **multi-purpose space stations**—not because computing naturally requires hotels in space, but because several expensive infrastructure requirements overlap.

This idea already has a primitive precedent in the International Space Station. The ISS is not a data centre, but it demonstrates that humans can continuously occupy a large orbital structure when it has reliable power, communications, environmental control, resupply and maintenance. Its solar arrays generate electricity; life-support systems maintain a habitable environment; radiators reject waste heat; communications connect the station with Earth; spacecraft dock with it; and orbital manoeuvres help maintain its trajectory. A future computing station would have different priorities, but many of the engineering categories would be familiar. **Life support** means the systems that maintain breathable air, safe pressure, temperature and other conditions required for people to survive. A computing station that normally operates robotically might need only a small human-rated section rather than making the entire facility habitable.

That distinction is important because humans and computers have different requirements. Servers do not need oxygen, food, bathrooms or comfortable living temperatures, while humans need protection from radiation and a carefully controlled environment. Future stations could therefore be divided into zones: large machine areas containing computing and storage equipment, robotic maintenance areas, power and thermal systems, cargo sections, docking interfaces and comparatively small pressurised human modules. Engineers might visit for weeks or months during major upgrades while routine maintenance is performed by robots. The objective would be to minimise the expensive infrastructure required to support people without eliminating human access altogether.

From Communication Hop to Service Station

The intermediate nodes discussed earlier in this book could evolve in the same way. Initially, a hop might primarily receive, store and forward information. Later generations could gain more computing capacity, additional storage, navigation services and robotic repair systems. If spacecraft regularly travel through the same broad transport corridors, selected larger nodes could also provide docking and logistics support. A spacecraft might dock to transfer cargo, receive replacement components, obtain locally available propellant or supplies where technically possible, undergo robotic inspection, exchange large datasets or provide a temporary refuge for its crew.

The analogy is not a motorway service station in the literal sense, because spacecraft cannot simply stop beside a fixed object in space. Both the spacecraft and station are moving according to orbital mechanics. **Rendezvous** means carefully adjusting a spacecraft's trajectory and velocity so that it can meet another moving object. Docking therefore costs time and propellant and would occur only where the benefit justified it. This is why we should not imagine evenly spaced stations forming a straight road between Earth and Mars. Useful facilities would instead occupy carefully selected orbits and locations associated with major transport routes, planetary systems and perhaps some stable or operationally useful regions discussed earlier.

A hierarchy could eventually emerge: small, unmanned communication relays; larger storage-and-compute nodes; robotic logistics depots; and a smaller number of major human-capable stations. Not every network hop needs beds, food and docking ports. Building every node to the highest specification would be unnecessarily expensive. This resembles terrestrial cloud architecture: a tiny edge site and a hyperscale data centre both perform computing, but they are designed for completely different scales and purposes.

The Station Must Repair Itself

The farther a station is from Earth, the more important autonomous maintenance becomes. A terrestrial data centre can call engineers when a cooling pump fails. An orbital station may not have that option. Sensors would continuously monitor power systems, electronics, communications, structural components and thermal equipment. AI could identify abnormal behaviour and direct robots to inspect the affected equipment. Modular construction would make this easier: rather than asking a robot to repair a complex circuit board component by component, it might remove a failed module and install a replacement.

This is **predictive maintenance**—using measurements and historical behaviour to identify equipment likely to fail before the failure actually occurs. A pump showing increasing vibration, for example, could be replaced during planned maintenance rather than after it stops. AI becomes valuable because a large station could contain millions of measurements continuously describing its condition.

Self-repair, however, cannot guarantee survival. A severe collision, fire, power failure or other event might disable an entire station. That is why the zonal redundancy described earlier remains essential. Another appropriately positioned facility should be capable of assuming important communication or computing functions while repair spacecraft or robots are dispatched. Some failures will be repairable; others may require abandoning the damaged infrastructure. **Resilience means designing the network to survive the loss of components, not pretending components will never be lost.**

Humans Become Visitors Rather Than Permanent Operators

Early extraterrestrial infrastructure may require substantial human involvement, but increasing autonomy could reverse the relationship familiar from Earth. Instead of people permanently operating the machines, AI and robots operate the station while people visit when necessary. A specialist team could arrive for a major hardware expansion, scientific mission or complex repair and then leave again. Other facilities, particularly around the Moon or eventually Mars, might support longer human stays.

Those living quarters could also provide emergency capability for travellers. A spacecraft experiencing a manageable problem might use a suitable nearby station for communications, diagnostics, supplies or temporary shelter if orbital geometry and available resources made rendezvous practical. We should not overstate this possibility—space is enormous and a station cannot rescue a spacecraft hundreds of millions of kilometres away merely because both are described as being "between Earth and Mars." But as traffic increases, strategically located infrastructure could make some routes safer.

A Network Begins Becoming a Transport System

This is how the Interplanetary Cloud could expand beyond its original digital purpose. Communication requires power. Power infrastructure can support computing. Computing supports autonomous robots. Robots maintain equipment. Maintenance creates demand for spare parts and logistics. Logistics creates docking infrastructure. Human maintenance creates demand for accommodation and life support. More spacecraft create demand for navigation, communications and emergency services.

One infrastructure layer creates a reason to build the next.

The transformation would therefore probably be gradual:

Relay → Storage Node → Compute Node → Robotic Service Hub → Logistics Station → Human-Capable Space Station

Not every facility follows the entire path, and such infrastructure would take decades or centuries to develop. But this is how large technological systems often evolve: facilities built for one purpose become platforms supporting functions their original designers did not require.

A data centre in space therefore does not have to remain merely a room full of computers floating above Earth. At sufficient scale, it could become part of the physical infrastructure through which spacecraft, robots, humans and information move around the Solar System.

And that produces another opportunity. Once we possess powerful computing facilities far from Earth's population and ecosystems, some of those facilities could provide something Earth cannot easily offer: **remote environments in which increasingly capable AI and robotic systems can be tested under tightly controlled conditions.**

Chapter 29

Extraterrestrial AI Laboratories

Artificial intelligence and robotics will need greater autonomy as infrastructure moves farther from Earth. A lunar construction robot may sometimes receive rapid human assistance, but a Martian machine cannot depend on an immediate answer from Earth. It must perceive its surroundings, make approved local decisions, recover from some failures and recognise when a problem exceeds its authority.

That creates a legitimate safety question: where should increasingly capable autonomous physical systems be tested? Over the long term, selected extraterrestrial environments could provide useful physical sandboxes, but only if isolation is engineered rather than assumed.

Physical Distance Is Not Automatic Safety

Placing an experimental AI on the Moon or Mars does not contain it if the system has unrestricted communications, can issue commands to terrestrial equipment or controls other valuable infrastructure. Safety depends on permissions, interfaces, monitoring, software isolation and governance.

A true air gap is physical separation between a computer system and untrusted networks. Even then, an air gap is only one control. Physical access, removable media, authorised data transfers and connected machinery can still create pathways for failure.

The Moon Could Test Autonomous Robotics

The Moon is interesting because it combines a genuinely difficult environment with relative proximity to Earth. Future robots could excavate material, transport cargo, assemble solar arrays, inspect structures and maintain communications while dealing with real dust, vacuum, radiation, temperature variation and unfamiliar terrain.

Instead of remotely controlling every movement, humans could provide a bounded objective such as preparing a site for a solar array. Survey, excavation, transport, assembly and inspection robots would coordinate locally while humans define the mission limits and review exceptions.

Failure Can Become Useful Information

In a properly isolated test zone, a robotic failure can become an engineering lesson rather than a public disaster. If a construction machine damages an unfinished structure, engineers can analyse the event, improve the model or control system and test again. The purpose is not to make failure acceptable; it is to create conditions in which learning from failure does not unnecessarily endanger populated areas.

The same principle could apply to selected high-energy industrial experiments, new autonomous maintenance systems or robotic manufacturing techniques. But scientific sites, operational infrastructure and future settlements still require protection.

Autonomy Needs an Operating Envelope

A trustworthy autonomous system should know the range of conditions in which it is authorised to act. Inside that operating envelope, it can make local decisions. Outside it, the system should move toward a safe state, isolate affected equipment and request assistance rather than improvising without limit.

This distinction is central to the Interplanetary Cloud: autonomy is necessary because distance prevents continuous control, but autonomy should remain bounded by policy, safety rules and explicit authority.

Planetary Air-Gapping Could Add One Layer

For some future experiments, physical separation could become an additional defence. A remote facility might use narrow, rate-limited communication channels, independent validation gateways and no direct authority over terrestrial systems. Outputs could be reviewed before being forwarded to wider networks.

This does not turn space into a dumping ground for dangerous technology. The more powerful the experiment, the more independent barriers should exist between a mistake and a catastrophic consequence. Safety must grow with capability.

The Main Value Is Responsible Learning

The strongest argument for extraterrestrial laboratories is therefore not that dangerous AI should be moved away from Earth. It is that a civilisation operating increasingly autonomous machines will need realistic environments in which those systems can be tested, observed and improved under strict controls.

The lessons would return to Earth as well. Better fault detection, bounded autonomy, robotic recovery and safe-mode design developed for space could improve terrestrial infrastructure. The laboratory is valuable only if the knowledge gained makes systems safer everywhere.

PART VII THE SPACE DATA-CENTRE ECONOMY

From technical infrastructure to work, industry and governance

Closing Part VI

Part VI transformed the Interplanetary Cloud from a network of hypothetical data centres into a wider industrial ecosystem. Computing depends on energy, communications, logistics, robotics, resources, maintenance and, in selected locations, human presence. Once these systems begin supporting one another, they no longer represent isolated engineering projects. They begin to form an economy. That creates two questions that technology alone cannot answer: who participates in that economy, and who controls the infrastructure on which it depends?

What This Part Covers

The infrastructure described so far would require far more than cloud engineers and rockets. It would create demand across energy, robotics, communications, cybersecurity, logistics, manufacturing, mining, construction, medicine and many other fields. This Part examines how an extraterrestrial computing ecosystem could therefore become a wider economy and reconnects that possibility with the employment disruption that opened the book. It then addresses governance. A distributed interplanetary infrastructure should not become a technological empire controlled by one company, nation or AI system. Instead, independently operated systems could cooperate through shared standards, federated governance, interoperability and institutional redundancy, while AI manages complexity within boundaries established by human institutions.

Chapter 30

The Space DataCentre Economy

The previous chapters described robots constructing infrastructure, orbital data centres becoming multi-purpose service hubs, and extraterrestrial locations supporting increasingly autonomous machines. If even part of this future develops, it creates something larger than a technology project: a new economy.

Data centres beyond Earth would need processors, rockets, power systems, radiators, communications, robotics, cybersecurity, mining, manufacturing, logistics and maintenance. If people work at selected facilities, life support, medicine, habitats and transport add further layers. Each requirement creates demand for skills, suppliers and supporting infrastructure. That matters because this book began with the opposite problem: AI disrupting employment on Earth.

Space infrastructure cannot solve today's employment disruption on its own. Large lunar compute regions will not appear next year, and Martian industry will not absorb workers who lose jobs next month. But over a much longer horizon, the same technology that automates some existing work could help create industries that do not yet exist.

AI May Automate Tasks Faster Than Professions

The employment debate often treats a job as one indivisible unit: either AI replaces it or it does not. Real work is more complicated. A software engineer writes code but also interprets requirements, reviews architecture, investigates failures and communicates with people. A platform engineer automates infrastructure but also manages risk, handles incidents and decides what should happen when automation fails. Similar mixtures exist across administrative, financial, legal, creative and customer-service work. AI can increasingly perform some of these tasks, which may allow fewer people to produce the same output. That transition can still be painful. Jobs can disappear, entry-level opportunities can shrink and workers may need to retrain. The long-term question is therefore not merely how many tasks AI automates, but whether society creates new objectives and industries large enough to redirect human capability.

A Lunar Compute Region Needs Far More Than IT

Consider one large lunar compute region. Processor and server specialists design the hardware. Power engineers provide electricity. Thermal engineers remove waste heat. Network engineers connect the region with Earth and orbital infrastructure. Robotics engineers build machines capable of inspection and repair. Cybersecurity teams protect the systems. Logistics organisations deliver equipment, while mining and materials specialists may eventually provide some local resources.

The data centre can therefore become an anchor customer: a major facility whose demand helps justify other infrastructure. We have seen versions of this process on Earth. Ports create logistics ecosystems. Airports attract transport and service industries. Mining projects create roads, energy systems and settlements. A large extraterrestrial compute project could have similar spillover effects, although at far greater cost and over a much longer timescale.

Existing Skills Could Evolve Rather Than Disappear

Take a platform engineer. AI may increasingly generate Terraform, deployment configurations, tests and troubleshooting suggestions. But an interplanetary environment creates architectural problems that are larger than routine configuration: which services must remain local to Mars, what happens when a planetary link disappears, how should workloads fail over, and which AI agents may perform automatic recovery? AI may write more of the implementation while the human increasingly defines boundaries, resilience and acceptable risk.

Software engineering would evolve similarly. Extraterrestrial systems cannot assume permanent Internet connectivity or immediate human assistance. A lunar mining robot, orbital power controller or Martian life-support system needs software designed to fail safely. Network engineers would work with optical links, moving nodes and Delay-Tolerant Networking. Cybersecurity specialists would protect spacecraft and robotic facilities in which a compromised digital system could produce physical damage.

The transformation would not be limited to white-collar technology work. Mechanical technicians could supervise robotic diagnostics. A geologist on Earth could analyse lunar mining data. Electricians and power engineers could design systems installed by robots. Much of the supporting workforce could remain on Earth: the physical infrastructure moves outward while a large part of the knowledge economy stays here.

New Specializations Will Grow From Old Ones

Some future job titles may sound unfamiliar: planetary cloud architect, interplanetary data-resilience engineer, robotic fleet reliability engineer, space-compute thermal engineer, orbital logistics optimiser or planetary AI safety engineer. The exact titles are impossible to predict. The important point is that new infrastructure creates new specialisations.

Many roles would combine disciplines that are currently more separate. Computing may blend with robotics; networking with orbital mechanics; geology with AI; cybersecurity with autonomous systems. A worker may become less defined by one narrow tool and more by the system they are responsible for keeping safe and useful.

This does not mean every displaced office worker can simply become a spacecraft engineer. Some roles will demand advanced technical education; others will emerge in construction, manufacturing, operations, finance, procurement, law, safety, training and project management. A future space economy would extend far beyond astronauts and rocket scientists.

AI Could Help Workers Make the Transition

There is an important circular relationship here. AI contributes to disruption, but it can also lower some barriers to retraining. A network engineer could use AI to learn optical space communications. A software engineer could move toward robotics. A mechanical technician could learn autonomous diagnostics. A graduate could practise with simulations that once required expensive equipment.

The durable skill may increasingly become the ability to extend existing experience into a new technical environment. Human knowledge is not discarded; it becomes the foundation for the next specialisation.

The Employment Bridge Still Starts on Earth

None of this removes the near-term responsibility described earlier in the book. Workers cannot be told simply to wait for industries that may take decades to mature. Governments, companies and educational institutions need practical routes into expanding sectors now: energy, robotics, cybersecurity, advanced manufacturing, resilient infrastructure and other areas that support both terrestrial needs and a possible space future.

The space economy is one possible destination. The Employment Bridge must be built long before we arrive there.

Computing Could Become an Economic Anchor

Why might computing itself become an anchor rather than simply following settlements? Because autonomous machines, communications and local decision-making may be required before large human populations exist off Earth. Robots surveying terrain, operating mines, maintaining power systems and constructing infrastructure all need computation. A compute facility could therefore arrive alongside, or even before, substantial settlement.

Once it exists, the infrastructure starts sharing itself. A power installation built for computing may serve nearby industrial equipment. Communication links can carry scientific traffic. Robotic maintenance systems can service mining machinery. Landing facilities built to receive server hardware can accept other cargo. If people are present, water processing, habitats and medical support can serve a wider community.

The progression might look like this: Compute → Power → Communications → Logistics → Resource Extraction → Manufacturing → Human Services → Wider Settlement. This is not a prediction that every lunar settlement would grow around a data centre. It illustrates how one valuable piece of infrastructure can make the next investment easier to justify.

Mining, Manufacturing and Digital Inventory

Local resource utilisation could strengthen the economic loop. Early facilities would depend heavily on Earth for advanced electronics and specialised machinery. Over time, however, locally available materials might provide structures, shielding, roads, landing surfaces or selected mechanical components. Earth could continue supplying high-complexity technology while off-world industry increasingly supplies high-mass basic material.

Manufacturing then turns data into physical capability. Instead of shipping every simple replacement component from Earth, a verified design file could be transmitted to a local manufacturing system. A damaged bracket, housing or pipe might be produced near where it is needed. The data centre therefore becomes part computer facility, part library and part digital warehouse.

This does not eliminate conventional logistics. Advanced processors, precision electronics and many specialised parts could remain difficult to manufacture locally for a long time. The useful principle is narrower: transport information instead of mass whenever the economics, quality requirements and available manufacturing capability make that sensible.

Energy, Logistics and Clusters

Energy would likely become one of the first shared commodities. A compute facility may be able to reduce delay-tolerant workloads when a settlement or industrial system needs priority. The data centre becomes a participant in an intelligent local microgrid rather than simply a fixed consumer.

Regular cargo traffic can justify better landing facilities, orbital transfer hubs and inventory systems. Those investments then reduce the cost of supporting additional activities. Related organisations may gradually form clusters that share workers, suppliers, power, communications and repair infrastructure. The first facility is extraordinarily expensive; later participants can benefit from what already exists.

Earth Remains Part of the Space Economy

Moving infrastructure outward does not mean moving all economic benefit away from Earth. For a very long time, Earth would remain the centre of advanced manufacturing, research, finance, education and human population. A lunar compute project could create terrestrial demand for processors, rockets, robotics, software, communications equipment, energy technology and engineering services.

A technician may never leave their home city yet spend an entire career supporting extraterrestrial infrastructure. The space economy does not create jobs only in space. It creates jobs because of space.

The First Projects May Not Be Commercially Profitable

Early extraterrestrial infrastructure may not make conventional commercial sense. Strategic infrastructure often creates wider benefits that are larger than the direct revenue collected by its operator. Governments may fund research and foundational infrastructure; universities may develop technologies; private companies may commercialise successful components; international organisations may establish standards.

As launch costs fall, experience accumulates and supporting infrastructure grows, more activities could become commercially viable. The economic anchor may begin as a scientific or strategic project before becoming part of a broader market.

Growth Must Be Useful and Sustainable

Economic expansion also creates risks. Competition for favourable sites, orbital positions, communication capacity or extraterrestrial resources could reproduce destructive terrestrial behaviour. Development therefore needs rules for safety, scientific preservation, resource use and conflict prevention.

A reinforcing loop could emerge: AI improves robotics → robotics reduces infrastructure cost → cheaper infrastructure expands computing → expanding computing improves AI, manufacturing and logistics → new industries become possible. But the objective cannot be infinite computation simply because more computation is possible. The loop is valuable only if the expansion remains useful and sustainable.

From Job Disruption to Civilisation Expansion

The book began with a worker asking what AI means for a career. That question remains legitimate. Promising that technology may create future industries does not pay today's mortgage or remove today's uncertainty.

But the long-term response to automation cannot be preserving every existing task forever. Civilisation can create new objectives large enough to use the capabilities automation releases: cleaner energy, climate adaptation, advanced medicine, environmental restoration, scientific discovery and eventually infrastructure beyond Earth.

In that future, AI becomes more than a competitor for existing work. It becomes an amplifier of what humanity is capable of building. That opportunity is not guaranteed, and it will not arrive automatically. It must be created through investment, education, infrastructure and institutions.

Chapter 31

Who Controls the Interplanetary Cloud?

Once extraterrestrial computing becomes an economy rather than a collection of isolated machines, the hardest questions may no longer be technical. Who owns a lunar data centre? Who controls an orbital communication hub? Who receives access to scarce power, landing areas or communication capacity? And what happens when infrastructure built by one organisation becomes essential to people and machines operated by many others?

The Interplanetary Cloud should not depend on one company, one country or one autonomous AI possessing unrestricted control. Its technical architecture is distributed; its governance will probably need to become distributed as well.

A Federated System Is More Plausible Than One Owner

The Internet provides a useful broad analogy. No single organisation owns every network, data centre, cable and service, yet common protocols allow independently operated systems to communicate. An interplanetary infrastructure could develop in the same general direction.

One organisation might operate lunar compute facilities, another an Earth-Moon optical network, another orbital logistics stations and another Martian storage. Their equipment can remain independently owned while common standards allow data, spacecraft and services to move between them. Interoperability means independently built systems can cooperate rather than becoming isolated technological islands.

Ownership and Governance Are Different

A company may own servers without gaining the right to make laws for everyone using the Moon. A nation may finance a communication station without automatically gaining authority over every spacecraft that depends on it. Shared infrastructure may therefore require consortiums or agreements among organisations that remain independent.

Cloud providers, aerospace companies, telecommunications operators, energy companies, mining organisations, universities and governments each bring different capabilities. No cloud company needs to become a rocket company, mining company, electricity authority and planetary government at the same time.

Avoid Single Points of Control

If one organisation controls launch, communications, power, data centres, identity systems and robotic infrastructure across an entire planetary region, others become dangerously dependent on it. The same logic that discourages single points of technical failure also applies to economic and political control.

Multiple communication providers, compute operators and recovery systems create resilience. A cyberattack, commercial dispute, bankruptcy or policy decision by one organisation should not be capable of disconnecting an entire world.

Technical Interoperability Can Survive Political Difference

Governments will remain deeply involved because space infrastructure has strategic importance. Political competition will not disappear. But political independence does not require technical incompatibility.

Competing countries can still agree on enough common standards for emergency communications, navigation, docking and basic data exchange. Otherwise civilisation may waste enormous resources building duplicate systems solely because independently developed infrastructure cannot interoperate.

AI Can Operate the System Without Governing It

The Interplanetary Cloud could eventually become too complicated for humans to manually manage every server, power system, robot, communication route and spacecraft connection. AI can route data around failures, schedule workloads according to energy availability, predict equipment faults, manage storage and dispatch repair robots.

But operational intelligence should not quietly become political authority. There is a fundamental difference between asking AI how to achieve an approved objective and allowing AI to decide what objectives civilisation should pursue.

An AI scheduler may determine where a non-critical workload can run efficiently. It should not independently decide that a human settlement deserves less electricity because model training produces a higher calculated return. Humans establish laws, rights, priorities and safety boundaries; AI handles complexity inside them.

There should also be no single master AI controlling the entire system. Specialised and independently governed systems can supervise power, communications, life support and security while checking one another. Critical infrastructure should not trust one intelligence without limit.

Identity and Access Become Interplanetary Problems

Governance also appears at the level of individual permissions: who is allowed to do what? Future identities may include people, spacecraft, robots, AI agents, data centres and autonomous industrial systems. A lunar mining robot should not gain permission to control a habitat's oxygen system simply because both are connected to the same network.

Least privilege means each identity receives only the permissions required for its authorised function. A relay may forward encrypted information without being able to read it. A scientific AI may analyse geology without controlling mining machinery. Emergency systems may receive additional authority only under predefined conditions.

Each major planetary region also needs enough local identity and security infrastructure to continue operating when Earth is unreachable. A Mars settlement cannot become unusable because an authentication service on Earth is temporarily unavailable.

Data and Resource Governance Will Be Different Problems

Data sovereignty becomes more complicated when information crosses worlds and organisations. A medical record created on Mars might be stored locally and backed up elsewhere. Personal information requires privacy, scientific knowledge may benefit from broader access, commercial information requires intellectual-property protection, and civilisation archives may require safeguards against deletion by any one actor.

Physical resources raise different questions. Water-bearing regions, useful orbital locations, communication frequencies and energy-rich sites could become economically important. Governance must balance incentives to invest with common interests, scientific preservation and the need to avoid permanent monopolies over essential infrastructure.

Institutional Redundancy Matters

A resilient network should assume that governments change, companies fail, contracts end and international relations sometimes deteriorate. Critical infrastructure should not require permanent political harmony to remain safe.

Common emergency standards, alternative communication routes, independent life-support capability and archives operated by multiple authorities create institutional redundancy. Engineers build redundancy because machines fail. Governance needs redundancy because organisations can fail too.

Governance Will Evolve With Distance

The Moon could initially remain closely connected to institutions on Earth because communication is rapid. Mars is different. A settlement millions of kilometres away cannot realistically ask Earth for permission for every operational decision. As populations and economies develop, greater local governance would become increasingly necessary.

The political structure may therefore mirror the technical one: local decisions where delay makes central control impractical, shared standards connecting regions, and interplanetary institutions coordinating issues that affect everyone.

The Interplanetary Cloud Should Be an Ecosystem, Not an Empire

So who controls the Interplanetary Cloud? The best long-term answer may be: no one controls all of it.

Companies can own and operate infrastructure. Governments regulate activities and protect public interests. International agreements establish common rules. Technical standards allow independent systems to cooperate. Local settlements eventually gain responsibility for local affairs. AI manages complex operations within human-defined boundaries. Independent institutions provide checks against excessive concentration of power.

That arrangement is less simple than one central authority. But distributed systems teach an important lesson: centralisation can make management easier while making failure more dangerous. For infrastructure supporting multiple worlds, political resilience may become as important as server redundancy.

PART VIII TOWARD AN ADVANCED CIVILISATION

What it means for biological and artificial intelligence to grow together

Closing Part VII

Part VII showed that technical decentralisation eventually requires economic and institutional decentralisation as well. Resilience cannot apply only to servers and communication links; it must also apply to ownership, governance and authority. The Interplanetary Cloud works best as an ecosystem rather than an empire. This brings the book to its final question. If humanity eventually possesses the intelligence, automation and infrastructure required to operate across worlds, technological capability alone will not make us an advanced civilisation. What matters is how we choose to use it.

What PART VIII Covers

The final Part brings the argument back to humanity. Technological sophistication is not the same as civilisational advancement. AI may become extraordinarily capable, robots may operate where humans cannot, and infrastructure may eventually extend across the Solar System, but those achievements matter only if they improve rather than diminish civilisation. These chapters examine responsible acceleration, environmental stewardship, meaningful human participation, cooperation, distributed governance and the complementary strengths of humans, AI and machines. Earth remains the foundation, not a disposable starting point. The ultimate purpose of the Interplanetary Cloud is therefore not simply to build data centres in space. It is to explore how greater intelligence could allow civilisation to expand its possibilities without abandoning responsibility for the world and people from which that future begins.

Chapter 32

The Future We Have to Earn

Humanity has achieved extraordinary things. We have split the atom, mapped the human genome, built spacecraft that have travelled beyond the outer planets, connected billions of people through global networks and created artificial intelligence capable of assisting with language, software, images and scientific analysis. By comparison with civilisation only a few centuries ago, we are unquestionably technologically sophisticated.

But technological intelligence and an advanced civilisation are not necessarily the same thing. A civilisation should perhaps be judged not only by what it can invent, but by whether it can manage the consequences of its inventions. That question connects the entire book. AI is becoming more capable while societies still struggle with unequal access to opportunity, environmental pressure, technological disruption and coordination across institutions. Our engineering capability can advance faster than our ability to decide what that capability is for.

Intelligence Solves Problems; Advancement Avoids Creating Larger Ones

Intelligence can produce powerful technology. Advancement requires deciding when, where and how that technology should be used. An intelligent civilisation may discover a new source of energy. An advanced civilisation also considers safety, environmental consequences, accessibility and long-term effects.

AI can automate thousands of repetitive tasks. Advancement requires ensuring that productivity gains do not simply leave millions of people without meaningful economic participation. This does not require technological progress to stop. The challenge is responsible acceleration: developing capability while energy systems, regulation, education, employment and social institutions evolve alongside it.

Earth Is Our First Test

Before managing resources across the Solar System, humanity still has to improve how it manages one planet. Access to food, water, electricity, healthcare, education and digital infrastructure remains unequal even when the underlying technology exists.

Climate change is a collective-action problem: many individually rational decisions can combine into a damaging global result. AI infrastructure

creates a smaller version of the same challenge. Individual companies and countries have incentives to expand computing, while the combined demand can place pressure on electricity networks, land, water and emissions depending on how the infrastructure is built.
An advanced civilisation needs to recognise system-level effects before they become crises.

AI Is a Test of Civilisational Maturity

If AI raises productivity dramatically, who benefits? If it reduces the need for some occupations, how do people transition? If autonomous systems control critical infrastructure, who remains accountable? If AI requires enormous computing resources, how is that energy supplied sustainably? If a small number of organisations possess the most powerful systems, how do we avoid excessive concentration of power?
These are not arguments against AI. They are reasons to take its success seriously. A technology powerful enough to transform civilisation requires civilisation to transform around it.

The Goal Cannot Be Infinite Consumption

Technological advancement does not mean consuming ever larger quantities of resources forever. More computing is valuable only when it produces sufficient benefit. Storing data nobody will use is not progress. Building infrastructure simply because capital is available is not progress. Extracting extraterrestrial resources without a useful purpose is not progress.
A mature civilisation should improve efficiency as well as expand. This is why earlier chapters considered workload placement, energy-aware computing, data deletion, local processing, repair and recycling. The objective is not to make the Interplanetary Cloud enormous. It is to make it useful.

Waste May Be a Measure of Immaturity

Space makes waste extraordinarily expensive. A Martian settlement cannot casually discard water. A lunar habitat cannot assume unlimited replacement material. A remote data centre cannot throw away functioning equipment simply because replacing it from Earth is convenient.
Scarcity encourages circular engineering: repair equipment, recover materials, recycle water, reuse components and design for long operational life. The lessons required to survive away from Earth may also teach us to use resources more intelligently on Earth.
An advanced civilisation may ultimately be identified less by how much it consumes than by how little it needs to waste.

Humans and AI Do Not Need the Same Physical Habitat

One important implication of this architecture is that biological and artificial intelligence do not need identical environments. Humans need breathable air, water, food, pressure, radiation protection and comfortable temperatures. Data centres and robots require power, communications, thermal control and maintenance, but they do not need forests, farmland or cities designed around human biology.

Over decades and centuries, this could allow some resource-intensive machine infrastructure to expand into environments where permanent human settlement is unnecessary. Earth remains overwhelmingly biological. Selected lunar and orbital regions become machine-oriented industrial habitats. Mars may eventually contain both human settlements and extensive autonomous industry. More hostile destinations could remain predominantly robotic.

This is not separation between humans and AI. It is a division of capability. Humans provide purpose, values, judgement and governance. AI provides computation and analysis. Robots provide physical reach where biological bodies struggle.

AI Needs Humanity Too

The relationship is not one-directional. AI does not independently manufacture processors, launch rockets, establish laws or decide why an infrastructure project is worth building. Human civilisation provides the physical, economic and institutional system in which AI exists.

The better objective is therefore co-evolution: biological civilisation expands its capabilities using artificial intelligence, while AI develops within institutions that keep its growth connected to human purposes.

Earth Becomes More Valuable, Not Less

The ability to place some infrastructure elsewhere should increase our appreciation of Earth. A server can theoretically operate in many environments if enough engineering surrounds it. Human beings cannot. Earth provides a functioning biosphere that no nearby world naturally replaces.

A lunar data centre can operate without forests. An asteroid robot does not need farmland. An orbital platform does not require rivers. Humans do. Moving appropriate infrastructure beyond Earth should therefore reduce pressure on our home, not reduce our responsibility toward it. The full lifecycle still matters: launches, manufacturing and off-world systems must create a genuine net benefit rather than merely relocating environmental costs.

We Are Not Advanced If Our Machines Advance Without Us

Imagine AI becoming extraordinarily capable while wealth and opportunity concentrate among a shrinking number of infrastructure owners. Technology may have advanced, but the civilisation around it may not have advanced at the same pace.

The purpose of automation should be to reduce unnecessary labour while expanding what people can accomplish, not to make human beings economically irrelevant. If AI releases human capability from repetitive work, society needs new objectives worthy of that capability: scientific discovery, medicine, education, resilient infrastructure, environmental restoration and, eventually, expansion beyond Earth.

The answer to automation is not inventing artificial tasks to keep people occupied. It is expanding the frontier of meaningful human activity.

An Advanced Civilisation Thinks Beyond One Generation

Interplanetary infrastructure forces a longer time horizon. A lunar industrial ecosystem could take decades to mature. A self-sustaining Martian settlement could require generations. A Solar System communications network could develop over centuries. Civilisation-scale archives might be designed to outlive the organisations that created them.

Intergenerational planning means making some decisions whose primary beneficiaries may not yet exist. That does not mean sacrificing the present for a speculative future. There is no useful distant future if we destroy the conditions required to reach it. Long-term planning begins by protecting the present.

Cooperation May Be Our Hardest Technology

Building processors, robots and spacecraft is difficult. Coordinating governments, companies, communities and competing interests may be harder still. The Interplanetary Cloud cannot realistically be built by one person, one company or one nation.

Standards, shared research, investment, regulation and long-term cooperation become part of the engineering problem. Cooperation itself becomes a form of civilisational technology.

An advanced civilisation may eventually be measured not only by how far its spacecraft travel, but by how many competing interests it can coordinate without destroying the objective they share.

What We Have to Earn

The future described in this book is deliberately ambitious. None of it is guaranteed. Some ideas may prove impractical, others may develop in forms very different from those imagined here, and many could take generations or centuries.

The deeper principles do not require us to wait. Use energy intelligently. Protect Earth. Develop AI responsibly. Help people adapt to automation. Design for resilience rather than unlimited consumption. Use robots where human biology creates unnecessary risk. Cooperate where problems are larger than individual organisations or countries. Preserve human accountability as machines become more autonomous.

We do not inherit an advanced civilisation simply because we invented artificial intelligence. It is a future we have to earn.

Chapter 33

Beyond the Cloud

For most of the digital era, the word cloud has been a useful abstraction. We open an application, upload a photograph, ask an AI a question or deploy software, and the infrastructure behind that action largely disappears from view. Yet the cloud has never actually been somewhere above us. It consists of physical data centres filled with processors, storage and networking equipment. Those facilities consume electricity, occupy land, require materials and generate heat.

Artificial intelligence is making this physical reality increasingly difficult to ignore. More capable models create demand for processors, electricity, cooling, storage and networks. What begins as a software revolution therefore becomes an infrastructure problem.

The answer is neither to stop AI nor to assume Earth must absorb every future stage of its physical growth. The long-term challenge is to find a path in which intelligence can expand while Earth remains protected as the biological foundation of civilisation.

The Cloud Becomes a Solar System Architecture

The first step beyond today's cloud does not need to be a gigantic Martian data centre. Expansion would probably be gradual. Terrestrial computing becomes cleaner and more efficient. Edge systems process more information close to where it is generated. Orbital infrastructure provides specialised communications, storage or computing. Lunar systems develop alongside scientific and industrial activity. Robots build and maintain increasingly complex facilities. Only much later might large compute regions emerge on the Moon, Mars or elsewhere.

The architecture could expand in layers: Terrestrial Cloud → Edge and Orbital Computing → Lunar Compute → Interplanetary Network → Planetary Compute Regions.

Each layer remains useful rather than replacing the one before it. Workloads run where their requirements make sense. Services needing immediate interaction stay close to users. Planetary infrastructure handles time-critical operations locally. Delay-tolerant work can move when energy, data location and capacity make another region advantageous.

Distance Does Not Disappear

No technological optimism removes the physics of distance. Communication with the Moon takes seconds for a round trip. Communication with Mars takes minutes. Intermediate hops cannot reduce the fundamental propagation delay because electromagnetic signals still cannot exceed the speed of light.

The consequence is architectural rather than merely inconvenient. Earth, the Moon, Mars, spacecraft and orbital stations become semi-independent compute regions capable of operating during temporary disconnection. Hops provide resilience, buffering and alternative routes; they do not make planets behave like racks inside one building.

The Interplanetary Cloud succeeds not by defeating distance, but by learning to operate within it.

Data, Machines and Infrastructure All Become Distributed

Data will travel continuously between worlds. Some information remains local, some is summarised, some is replicated for resilience and some is eventually deleted. Civilisation-scale archives extend disaster recovery beyond conventional geographic regions, but only selected information justifies the cost of deep planetary replication.

Communication infrastructure also becomes physical civilisation infrastructure. Selected nodes can combine communications, storage, computing, robotic servicing and, where justified, spacecraft docking or limited human accommodation. The boundary between data centre, network node and service station begins to blur.

AI gives this architecture physical reach. Engineers define objectives and safety limits. AI analyses conditions and coordinates operations. Robots inspect, construct and repair. Humans intervene where judgement, accountability or unusual circumstances require them.

The Definition of a Data Centre Changes

A future data centre may bear little resemblance to today's windowless terrestrial buildings. A lunar compute region might combine computing, storage, energy, communications, robotic maintenance, radiation protection and industrial support. A Martian facility might operate as part of a settlement. Orbital nodes may specialise in communication and caching.

The data centre stops being merely where computers are kept. It becomes part of the operating infrastructure of civilisation.

Earth Remains the Centre That Matters Most

Expansion beyond Earth must never make our current planet sound temporary. Earth possesses something no nearby destination naturally provides: a vast living environment in which human beings can survive without surrounding themselves with an artificial life-support machine. Mars does not provide that. The Moon does not. Orbital stations do not. Protecting Earth is therefore not simply one stage before moving somewhere else. It remains the biological foundation of the civilisation capable of making expansion possible.

Space should become another option when its advantages eventually justify its extraordinary engineering cost, not an excuse to avoid solving terrestrial problems.

The Future Is Not About More Servers

The central argument of this book is not simply to build enormous quantities of extraterrestrial computing. Efficiency matters. Cleaner grids matter. Better algorithms matter. Deleting unnecessary data matters. More efficient processors matter. Repair and reuse matter.

Moving inefficient infrastructure into space does not magically make it sustainable. The better future combines efficient terrestrial computing, cleaner energy and carefully justified extraterrestrial expansion.

From Competition With AI to Expansion With AI

At the beginning of the book, AI appears largely as a disruptive force. It generates code, automates analysis and performs tasks people expected to remain human work. By the end, the possibility is broader.

AI can assist engineers designing spacecraft, coordinate robotic construction, manage planetary compute regions, optimise power, identify useful materials and operate communication networks extending millions of kilometres. The same technology that reduces demand for some existing tasks can increase civilisation's ability to attempt entirely new ones.

That does not automatically solve unemployment. The transition must still be managed. But it changes the long-term objective from preserving every existing task to creating new frontiers of useful work while giving people realistic pathways into them.

Beyond the Cloud Means Beyond Our Current Assumptions

We assume data centres belong on Earth because almost all of them always have. We assume cloud regions are separated by kilometres rather than planets. We assume engineers can reach failed equipment, networks are

continuously connected, power comes from terrestrial grids and backups exist in another city or country.
Those assumptions are not laws of nature. They describe the architecture that makes sense at our present stage of civilisation.
A future compute region can belong to a planet. A network can tolerate minutes of latency. A robot can repair equipment without waiting for a person. A backup can exist on another world. A data centre can be designed primarily for machines. An engineer can operate infrastructure no human being has physically visited.

The Interplanetary Cloud

The Interplanetary Cloud therefore does not begin with launching enormous data centres into space. Its development would be a much longer progression:

More efficient computing on Earth → greater use of personal and edge AI → cleaner energy and responsible terrestrial infrastructure → AI and robotics applied to difficult physical work and infrastructure problems on Earth → increasingly capable autonomous machines → larger robotic industrial systems → controlled extraterrestrial industrial environments → use of local extraterrestrial energy and materials → orbital and lunar infrastructure → planetary compute and storage regions → interplanetary resilience.

At every stage, the previous layer remains important. Personal devices do not eliminate data centres. Space infrastructure does not eliminate terrestrial infrastructure. Robots do not necessarily eliminate humans. The architecture expands because civilisation's capabilities and activities expand.

Earth remains the biological centre that must be protected. It is where much of this technology is invented, where most people initially live and where the first generations of increasingly capable machines are developed. But Earth does not necessarily need to remain the physical location of every mine, machine, energy system, archive and computing facility that a much larger technological civilisation may eventually require.

That is the larger architectural idea behind this book: **protect the living planet, distribute intelligence where it works best, use machines where humans should not have to go, use local resources where transporting everything from Earth stops making sense, and**

gradually extend civilisation's infrastructure beyond the boundaries of one world.

Beyond the Cloud

Cloud computing taught us that computing capability did not need to sit on our desk. The next transformation may teach us that every data centre does not need to remain on our continent—and, much later, that every part of our digital infrastructure does not necessarily need to remain on our planet.

The cloud will not disappear. It may expand: from data centres to regions, from regions to planetary infrastructure, and perhaps eventually across worlds.

Whether that expansion represents progress will depend on what we do before it happens. If AI merely consumes more resources while increasing social disruption, moving some infrastructure into space will not make civilisation more advanced. But if we use intelligence to improve efficiency, protect Earth, expand knowledge, create opportunity and responsibly extend our capabilities, something more important will have happened.

We will have moved beyond the cloud not simply in distance, but in how we understand the relationship between intelligence, technology and civilisation.

The first cloud connected computers. The next connected regions. The Interplanetary Cloud could someday connect worlds. But the most important connection must remain the one between technological progress and human purpose.

Closing Part VIII / Closing the Journey

The book began with a personal and immediate question: what happens when AI starts performing work that people spent years learning to do? From there, the question expanded—from jobs to data centres, from data centres to energy and resources, from Earth to orbit, from orbit to the Moon and Mars, and eventually from technology to civilisation itself.

The Interplanetary Cloud is one possible architecture for that longer future. Its individual components may develop differently, arrive on different timescales or never become practical in the forms imagined here. The deeper proposition is simpler. Greater intelligence should expand human possibility rather than shrink it. Earth should be protected rather than exhausted. Machines should go where biology struggles. Humans should retain responsibility for the purposes those machines serve. And as civilisation gains greater technological power, its capacity for cooperation, restraint and long-term thinking must grow with it.

The future described in this book is not something technology will automatically give us.

It is something we would have to earn.

Selected Sources and Further Reading

The Interplanetary Cloud brings together established research, demonstrated technologies, emerging engineering capabilities and the author's own forward-looking architectural proposals. The sources below provide a selected research basis for factual claims and existing technologies discussed throughout the book. They are not intended to imply that the speculative architecture proposed in this book has already been demonstrated or endorsed by the organisations listed.

Artificial Intelligence, Employment and Software Engineering

International Labour Organization (2025), *Generative AI and Jobs: A 2025 Update*. International Labour Organization. Examines occupational exposure to generative AI and concludes that job transformation is currently more likely than widespread direct replacement.

Merola, R., Ernst, E., Samaan, D., del Rio-Chanona, M. and Teutloff, O. (2026), *The Impact of GenAI on Jobs, Productivity and Work Organization: A Review of the Empirical Evidence*. International Labour Organization. Reviews emerging evidence on productivity, employment and workplace organisation, including the limited evidence so far of large-scale displacement and possible pressure on opportunities for younger workers.

Anthropic (2025), *Anthropic Economic Index: AI's Impact on Software Development*. Analysis of 500,000 coding-related interactions across Claude.ai and Claude Code, including differences between automation and human-AI augmentation.

AI, Data Centres, Energy and Climate

International Energy Agency (2025), *Energy and AI*. IEA, Paris. Major assessment of data-centre electricity demand, AI-related energy growth, power infrastructure, emissions, energy security and the potential use of AI within energy systems.

International Energy Agency (2026), *Key Questions on Energy and AI.* IEA, Paris. Updated analysis of data-centre investment, electricity demand, grid constraints, energy security, affordability and sustainability.

Intergovernmental Panel on Climate Change (2023), *Climate Change 2023: AR6 Synthesis Report*. IPCC. Summarises the scientific assessment of human-caused climate change, observed warming, impacts and future risks.

Population and Long-Term Planning

United Nations Department of Economic and Social Affairs (2024), World Population Prospects 2024: Summary of Results. United Nations. Provides the projection that global population is expected to peak at around 10.3 billion in the mid-2080s before declining slightly to around 10.2 billion by 2100.

Space Communications and Delay-Tolerant Networking

Burleigh, S., Fall, K. and Birrane, E. (2022), *Bundle Protocol Version 7.* RFC 9171. Internet Engineering Task Force. Defines Bundle Protocol Version 7 for Delay/Disruption-Tolerant Networking in environments involving intermittent connectivity and large or variable communication delays.

NASA, *Laser Communications Relay Demonstration (LCRD)*. Overview of NASA's geosynchronous laser-communications relay demonstration and the use of optical communications to increase spacecraft data-transfer capacity.

NASA, Exploration and Space Communications: LCRD, section on the Integrated LCRD Low Earth Orbit User Modem and Amplifier Terminal (ILLUMA-T). Documents the International Space Station optical-communications demonstration with LCRD, including data transmission at rates of up to approximately 1.2 gigabits per second.

NASA, Laser Communications Relay Demonstration (LCRD) Overview. Includes NASA's High-Rate Delay/Disruption Tolerant Networking demonstration with LCRD and ILLUMA-T, including store-and-forward networking over high-rate optical links and its relevance to future interconnected space networks.

Orbital Mechanics, the Moon and Interplanetary Distance

NASA, *Moon Facts*. NASA Science. Provides basic lunar dimensions, average Earth-Moon distance, orbital behaviour and synchronous rotation.

NASA Jet Propulsion Laboratory, *Let's Go to Mars! Calculating Launch Windows*. Explains the moving-target problem in interplanetary travel, Hohmann-transfer principles and why spacecraft must launch when Earth and Mars have suitable relative positions.

NASA, *25 Years of Continuous Robotic Mars Exploration – From Pathfinder to Perseverance*. Includes discussion of the approximately 26-month cycle between favourable Mars launch opportunities.

NASA, *What Are Lagrange Points?* NASA Science. Explains the five Lagrange points created by the gravitational relationship between two major bodies and their usefulness for spacecraft positioning.

Spacecraft, Orbital Infrastructure and Thermal Control

NASA, *Reference Guide to the International Space Station*. Includes technical descriptions of the station's thermal-control systems, circulating coolant loops, heat exchangers and external radiators.

NASA Technical Reports Server, *International Space Station Active Thermal Control Sub-System On-Orbit Pump Performance and Reliability Using Liquid Ammonia as a Coolant*. Technical discussion of the ISS systems used to collect, transport and reject waste heat.

NASA, *Autonomous Landing and Hazard Avoidance Technology (ALHAT)*. Describes technologies for automated terrain assessment, hazard detection, navigation and precision landing for future planetary spacecraft.

Mars, Autonomous Exploration and Local Intelligence

NASA Jet Propulsion Laboratory (2021), *NASA's Self-Driving Perseverance Mars Rover "Takes the Wheel"*. Describes Perseverance's AutoNav system, including autonomous terrain mapping, hazard identification and route planning.

NASA Science (2021), *Driving Farther and Faster With Autonomous Navigation and Helicopter Scouting*. Discusses Perseverance's ability to perform navigation and route-selection functions locally rather than depending on continuous control from Earth.

NASA Jet Propulsion Laboratory, *Mars at a Glance*. Provides basic Martian orbital and environmental information, including atmospheric composition and the very low atmospheric pressure compared with Earth.

NASA Science, *Location of Large Subsurface Water-Ice Deposit in Utopia Planitia, Mars*. Describes radar observations indicating substantial subsurface water-ice deposits on Mars.

In-Situ Resource Utilization and Extraterrestrial Industry

NASA, *Overview: In-Situ Resource Utilization*. Reviews technologies and research intended to use locally available resources on the Moon and Mars rather than transporting every required material from Earth.

NASA (2023), *NASA's Oxygen-Generating Experiment MOXIE Completes Mars Mission*. Describes the Mars Oxygen In-Situ Resource Utilization Experiment aboard Perseverance and its successful production of oxygen from the carbon-dioxide-rich Martian atmosphere.

NASA Jet Propulsion Laboratory, *MOXIE Could Help Future Rockets Launch Off Mars*. Explains the solid-oxide electrolysis process used by MOXIE and the potential role of locally produced oxygen in future human exploration and propulsion.

NASA, *RESOURCE – Resources for Exploration & Science of OUR Cosmic Environment*. Research programme examining lunar volatiles, their availability and technologies for future lunar in-situ resource utilisation.

Space Governance and International Cooperation

United Nations Office for Outer Space Affairs (1967), *Treaty on Principles Governing the Activities of States in the Exploration and Use of Outer Space, Including the Moon and Other Celestial Bodies* (Outer Space Treaty). Provides the foundational international legal framework governing state activities in outer space.

A Note on the Book's Proposed Architecture

Concepts such as planetary compute regions, interplanetary disaster-recovery layers, orbital compute and storage hubs, Follow-the-Sun computing across worlds, AI-coordinated extraterrestrial industry, civilisation-scale interplanetary archives and the broader Interplanetary Cloud architecture are forward-looking proposals developed in this book. The sources above establish relevant scientific principles, existing technologies and emerging capabilities; they should not be interpreted as evidence that the complete architecture proposed here currently exists or is certain to be built.

www.ingramcontent.com/pod-product-compliance
Lightning Source LLC
LaVergne TN
LVHW010714061126
851857LV00014B/1184
* 9 7 8 1 7 6 4 8 9 3 8 2 4 *